THE CHEMISTRY BOOK

WORKBOOK

Nicholas Stansbie
Brett Steeples
Sarah Windsor

The Chemistry Book Units 3 & 4
1st Edition
Nicholas Stansbie
Brett Steeples
Dr Sarah Windsor
ISBN 9780170412476

Publisher: Rachel Ford
Editor: Kirstie Irwin
Proofreader: Marcia Bascombe
Production controller: Karen Young
Cover designer: Chris Starr (MakeWork)
Text designer: Leigh Ashforth (Watershed Art & Design)
Project designer: Justin Lim
Permissions researcher: Wendy Duncan
Typeset by: MPS Limited

Any URLs contained in this publication were checked for currency during the production process. Note, however, that the publisher cannot vouch for the ongoing currency of URLs.

For product information and technology assistance,
in Australia call **1300 790 853**;
in New Zealand call **0800 449 725**

For permission to use material from this text or product, please email
aust.permissions@cengage.com

ISBN 978 0 17 041247 6

Cengage Learning Australia
Level 7, 80 Dorcas Street
South Melbourne, Victoria Australia 3205

Cengage Learning New Zealand
Unit 4B Rosedale Office Park
331 Rosedale Road, Albany, North Shore 0632, NZ

For learning solutions, visit **cengage.com.au**

Printed in China by 1010 Printing International Limited.
1 2 3 4 5 6 7 23 22 21 20 19

CONTENTS

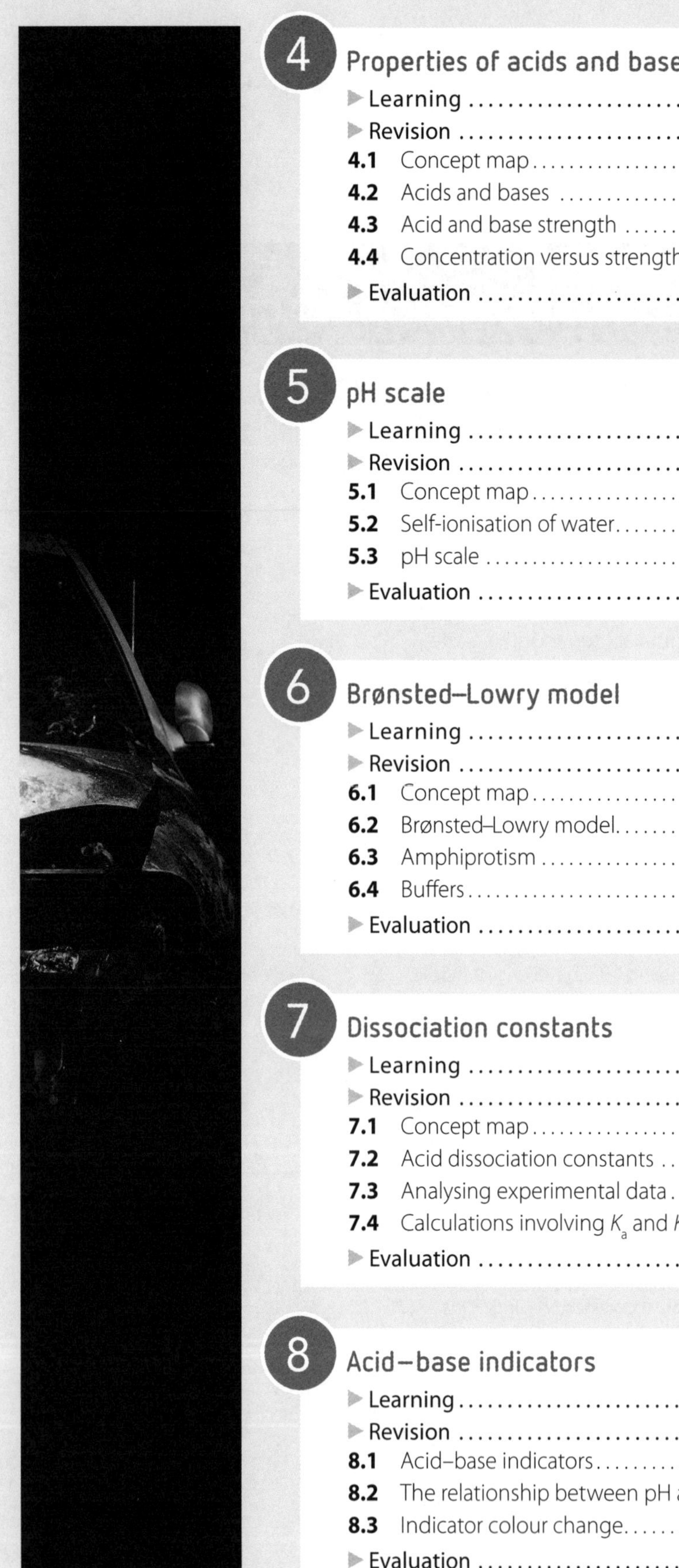

TOPIC 2: OXIDATION AND REDUCTION

TOPIC 1: PROPERTIES AND STRUCTURE OF ORGANIC MATERIALS

9780170412476

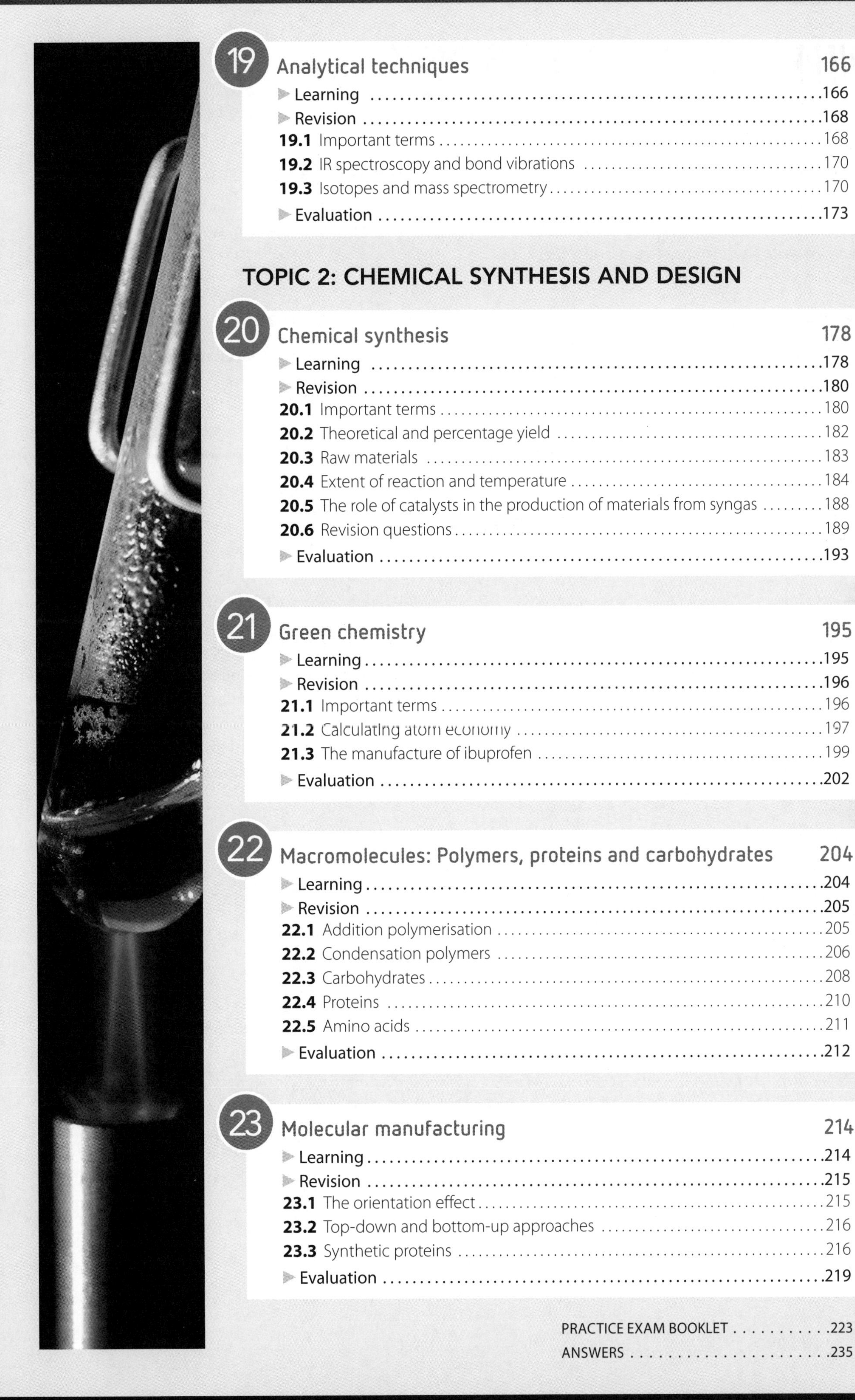

TOPIC 2: CHEMICAL SYNTHESIS AND DESIGN

HOW TO USE THIS BOOK

Learning

The learning section is a summary of the key knowledge and skills. This summary can be used to create mind maps, to write short summaries and as a check list.

Revision

This section is a series of structured activities to help consolidate the knowledge and skills acquired in class.

Evaluation

The evaluation section is in the style of a practice exam to test and evaluate the acquisition of knowledge and skills.

Practice examination

A tear-out exam helps to facilitate preparing and practising for external exams.

ABOUT THE AUTHORS

Nicholas Stansbie

Nicholas is an experienced science teacher and teacher of senior chemistry in the UK and in Australia. Nicholas's dedication to quality teaching and learning has enabled him to lead the *QScience Chemistry* team in the development of this text.

Brett Steeples

Brett was a research chemist and lecturer at Manchester University before becoming a chemistry teacher. Brett has worked on the moderation panel and is a member of the chemistry state review panel.

Dr Sarah Windsor

Dr Sarah Windsor is an associate lecturer in science at the University of the Sunshine Coast. Sarah's research interests include analytical chemistry and education and she has brought her extensive knowledge in both of those areas to the development of *QScience Chemistry*.

Some of the material in *The Chemistry Book Units 3 & 4* has been taken from or adapted from the following publications:

Nelson Chemistry for the Australian Curriculum Units 1 & 2 NelsonNet material written by: Deb Smith, Anna Davis, Anne Disney, Von Hayes, Rachel Whan, Suzanne Farr, Elizabeth McKenna and George Hook.

Nelson Chemistry for the Australian Curriculum Units 3 & 4 NelsonNet material written by: Bob Bucat and Rachel Whan.

SYLLABUS REFERENCE GRID

UNITS AND TOPICS	*THE CHEMISTRY BOOK UNITS 3 & 4*
UNIT 3: EQUILIBRIUM, ACIDS AND REDOX REACTIONS	
Topic 1: Chemical equilibrium systems	
Chemical equilibrium	Chapter 1
Factors that affect equilibrium	Chapter 2
Equilibrium constants	Chapter 3
Properties of acids and bases	Chapter 4
pH scale	Chapter 5
Brønsted–Lowry model	Chapter 6
Dissociation constants	Chapter 7
Acid–base indicators	Chapter 8
Volumetric analysis	Chapter 9
Topic 2: Oxidation and reduction	
Redox reactions	Chapter 10
Electrochemical cells	Chapter 11
Galvanic cells	Chapter 12
Standard electrode potential	Chapter 13
Electrolytic cells	Chapter 14
UNIT 4: STRUCTURE, SYNTHESIS AND DESIGN	
Topic 1: Properties and structure of organic materials	
Structure of organic compounds	Chapter 15
Physical properties and trends	Chapter 16
Organic reactions and reaction pathways	Chapter 17
Organic materials: Structure and function	Chapter 18
Analytical techniques	Chapter 19
Topic 2: Chemical synthesis and design	
Chemical synthesis	Chapter 20
Green chemistry	Chapter 21
Macromolecules: Polymers, proteins and carbohydrates	Chapter 22
Molecular manufacturing	Chapter 23

UNIT THREE

EQUILIBRIUM, ACIDS AND REDOX REACTIONS

- Topic 1: Chemical equilibrium systems
- Topic 2: Oxidation and reduction

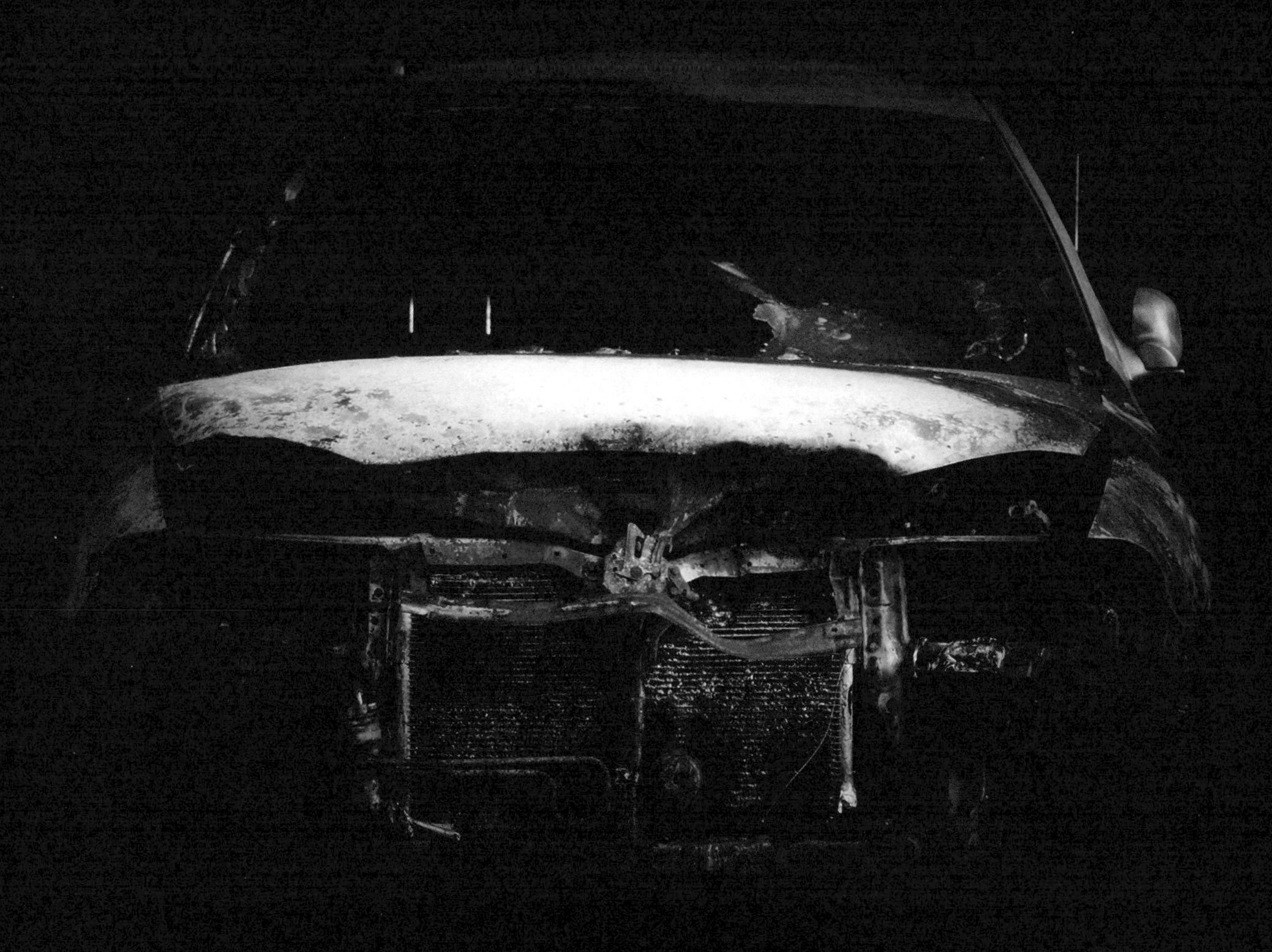

iStock.com/imamember

1 Chemical equilibrium

Summary

- An open chemical system is one where matter and energy can be exchanged with the surroundings; a closed system is one which is 'sealed', where matter and energy cannot be exchanged with the surroundings.
- Many chemical reactions are reversible. This means that as the reactants form products, the products can also react to reform the reactants.
- A reversible reaction can be represented by using the $\rightleftharpoons$ symbol in an equation. For example, the formation of ammonia from nitrogen and hydrogen is:

$$N_2(g) + 3H_2(g) \rightleftharpoons 2NH_3(g)$$

- In many cases, the forward and reverse reactions are occurring simultaneously. When both of these reactions occur at the same rate, there is no net change to the concentrations of the reactants and products, and the system is described as a dynamic equilibrium.
- Only some chemical changes are reversible. Whether or not a chemical reaction is reversible depends on the activation energies of the forward and reverse reactions, which must both be low in comparison to the energy the reacting particles possess, so that both reactions can occur easily.
- If we measure the concentrations of reactants and products in a reaction over time, we can identify when equilibrium has been reached and therefore how quickly it is reached.

1.1 Important terms

1 Complete the crossword using the clues on page 4.

using the clues on page 4

Across	**Down**
3 The chemicals involved in a reaction **4** A system in balance **7** A change involving the creation of a new substance **9** A __________ equilibrium is where forward and backward reactions are occurring at the same rate. **10** A change where no new substance is created **12** A system where one or more reactants or products in a reaction can be either added or lost	**1** Everything except the chemicals involved in a reaction **2** A reaction where the products can be converted back to the reactants **5** An __________ complex is a transition state, or intermediate formed during a reaction as a precursor to the product, but which has a high energy and so is relatively unstable. **6** A system where the chemicals involved in a reaction are all contained within a fixed space **8** Heat energy that is either absorbed or released in the breaking and making of chemical bonds during a reaction **11** The __________ energy is the minimum amount of energy required in order for a reaction to occur.

2 Write a short summary or create a mind map using all of the terms from the crossword.

1.2 Ethanoic acid and methanol in aqueous solution

Ethanoic acid and methanol in aqueous solution react slowly to form methyl acetate, reaching an equilibrium condition represented by the balanced chemical equation:

$$CH_3COOH(aq) + CH_3OH(aq) \rightleftharpoons CH_3COOCH_3(aq) + H_2O(l)$$

Methyl acetate belongs to a class of compounds called *esters*, formed by reaction between carboxylic acids and alcohols. This reaction, and corresponding ones to form other esters, is called *esterification.*

The reaction is catalysed by small amounts of sulfuric acid in solution. At any given time during the reaction, a sample can be taken from a reaction and the amount of ethanoic acid in the sample can be measured by titration with a standard sodium hydroxide solution, making a correction for the amount of sodium hydroxide solution that reacts with the sulfuric acid present.

Table 1.2.1 shows a typical set of measurements of the ethanoic acid concentration at various times throughout an experiment. The concentration of ethanoic acid is 1.00 mol L^{-1}, the concentration of methanol is 1.10 mol L^{-1} and the solution is at 25°C.

TABLE 1.2.1 The concentration of ethanoic acid over time

TIME (s)	INITIAL [CH_3COOH] (mol L^{-1})	CHANGE IN INTIAL [CH_3COOH] (mol L^{-1})	CHANGE IN [CH_3OH] (mol L^{-1})
0	1.00		
10	0.64		
20	0.46		
30	0.37		
40	0.30		
50	0.25		
75	0.17		
100	0.14		
150	0.10		
200	0.09		
250	0.09		

1 Use the grid below to plot the ethanoic acid concentration, with the concentration on the vertical axis ranging from 0–1.12 mol L^{-1} and the time on the horizontal axis ranging from 0–300 s.

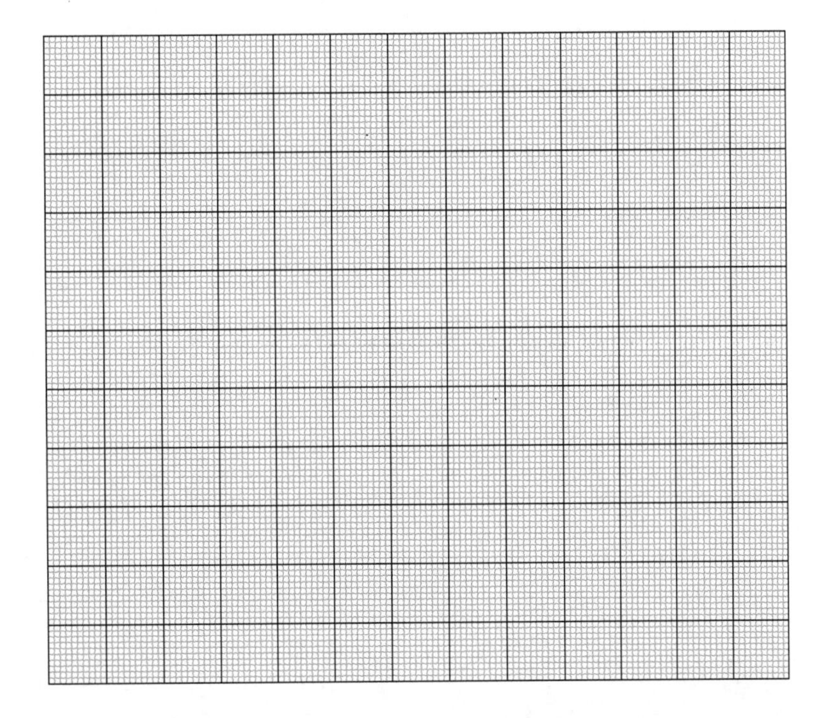

2 Why are more frequent measurements taken early in the reaction rather than later on in the reaction?

3 Use the balanced chemical equation for the reaction to deduce **the concentration of methanol** in each of the samples taken for ethanoic acid estimation. Record your results in Table 1.2.1 and add a plot of the change of ethanol concentration over time to the graph in question 1.

4 At approximately what time did the reaction mixture come to chemical equilibrium?

5 Was there any reaction occurring at $t = 250\,s$?

1.3 Forming stalactites and stalagmites

Research and investigate the chemistry of how stalactites and stalagmites are formed in limestone caves.

1 What is the role of carbon dioxide in:

a dissolving limestone as water seeps through it?

b deposition of calcium carbonate when the limestone-laden water enters a cave?

2 Discuss how both of these phenomena can be explained in terms of reversible reactions not at equilibrium.

9780170412476

The formation of ammonia gas from nitrogen and hydrogen can be represented by the equation:

$$N_2(g) + 3H_2(g) \rightleftharpoons 2NH_3(g)$$

When this reaction is carried out, the concentrations of nitrogen, hydrogen and ammonia gas are measured and represented on the graph below.

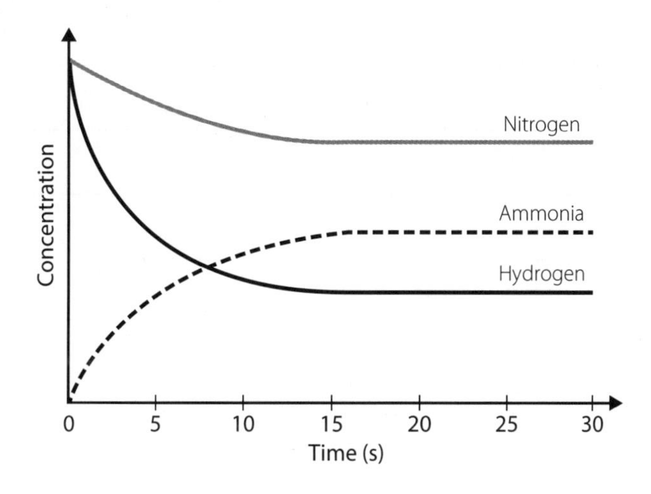

1 Explain what is meant by a dynamic equilibrium.

2 Explain why the equation contains the symbol $\rightleftharpoons$.

3 Once the reaction starts, explain why the hydrogen concentration falls more quickly than the nitrogen concentration.

4 An observer suggests that the reaction has stopped at 30 s. Explain why this is not the case.

2 Factors that affect equilibrium

Summary

- Once a system is at equilibrium, the concentration of reactants and products of the reaction remains constant.
- If the reaction conditions are changed, then the position of equilibrium can also change and become re-established with different constant values of reactant and product concentration.
- Changing temperature affects the position of equilibrium and the value of the equilibrium constant.
  - If the reaction is *endothermic*, the forward reaction is favoured by raising the temperature and the position of equilibrium moves to the right. The value of the equilibrium constant increases with temperature.
  - If the reaction is *exothermic*, the backward reaction is favoured by raising the temperature and the position of equilibrium moves to the left. The value of the equilibrium constant decreases with temperature.
  - The activation energy for an endothermic reaction is greater than that for an exothermic reaction. Therefore, when the temperature is raised, whilst the rate of both reactions increases, the rate of the endothermic reaction increases more than the rate of the exothermic reaction.
- Changing concentration affects the position of equilibrium.
  - Increasing the concentration of a reactant increases the rate of the forward reaction.
  - As the rate of the forward reaction increases, the concentration of products will subsequently increase; therefore, the rate of the backward reaction will increase until the rate of both reactions is equal and equilibrium is re-established.
- Changing volume and pressure affects the position of equilibrium.
  - Altering the total volume of a gaseous system causes a subsequent change in the concentration of all the reactants and products.
  - If there are different numbers of reactant and product molecules then this change in concentration will affect one of the reactions more than the other, thereby causing a change in the position of equilibrium.
  - A decrease in total volume (i.e. an increase in total pressure), favours the reaction where there are fewer product molecules than reactant molecules.
  - An increase in total volume (i.e. a decrease in total pressure), favours the reaction where there are more product molecules than reactant molecules.

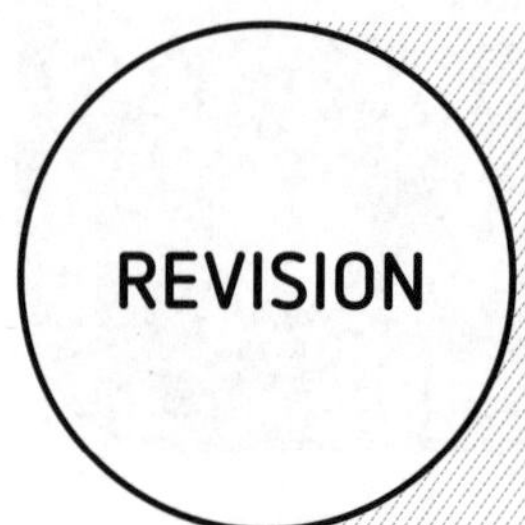

2.1 Important terms

1 Complete the crossword using the clues on page 11.

Across	Down
3 Le _______________'s principle states that if a change is imposed on a system at equilibrium, then the system will act in order to partially oppose the change.	**1** A substance that will alter the rate of a reaction without being consumed in the reaction
5 _______________ energy describes the minimum level of energy required in order for a reaction to take place.	**2** _______________ theory states that a reaction will occur when two particles collide with sufficient energy. The faster the collisions occur, the higher the rate of reaction.
6 Heat energy, stored in chemical bonds	**4** A reaction where the products have greater heat energy than the reactants, so that energy is absorbed from the environment when the reaction occurs, and the temperature falls
7 The amount of substance per unit volume	
8 A reaction where the products have lower heat energy than the reactants, so that energy is released to the environment when the reaction occurs, and the temperature rises	

2 Write a short summary or create a mind map using all of the terms from the crossword.

2.2 Concentrations of species

Figure 2.2.1 shows the change over time of concentrations of species in a reaction mixture that can be represented by the equation:

$$Fe^{3+}(aq) + SCN^{-}(aq) \rightleftharpoons FeSCN^{2+}(aq)$$

FIGURE 2.2.1 Graph of species concentration of a reaction mixture over time

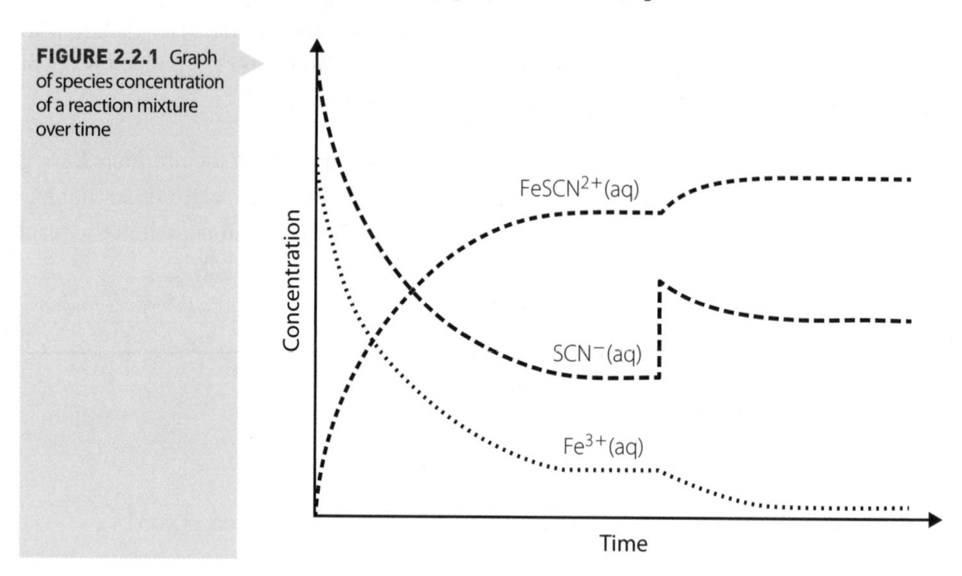

SCN⁻(aq) is the symbol for thiocyanate ions and $FeSCN^{2+}$(aq) represents a complex ion formed between iron ions and thiocyanate ions. Units are not given for either the concentration or the time axes.

The graph has three parts.

- The reaction mixture comes to equilibrium.
- A change is made to the system so that the reaction is no longer at equilibrium.
- The reaction mixture comes to a new condition of equilibrium.

Examine the graph and answer the following questions.

1 Which reagents were mixed initially? Which of the reactant species was present at the higher concentration?

____

____

2 Explain the changes, and the relative rates of change, in the concentrations of each of the species as the mixture comes to a condition of dynamic chemical equilibrium.

____

____

____

____

____

____

9780170412476

3 What change was made to the equilibrium reaction mixture so that it was no longer at equilibrium?

4 Explain the changes in the concentrations of each of the species as the mixture comes to a new condition of equilibrium after the change. Are the changes consistent with Le Chatelier's principle? Explain.

2.3 Boiling water

1 Explain why water boiling in a pot on the stove will not come to equilibrium with water vapour. Describe the conditions under which equilibrium between the liquid and vapour forms of water will come to equilibrium.

2.4 Explain using Le Chatelier's principle

Bromine, $Br_2(l)$, is not very soluble in water, but a dilute solution of it is brown. It is slightly acidic because of an equilibrium reaction represented by the equation:

$$Br_2(aq) + H_2O(1) \rightleftharpoons HOBr(aq) + H^+(aq) + Br^-(aq)$$

The dissolved species $Br_2(aq)$ is brown, while hypobromous acid, $HOBr(aq)$, and bromide ions, $Br^-(aq)$, are colourless.

1 Use Le Chatelier's principle to explain the observation that bromine water becomes colourless if sodium hydroxide solution is added but becomes brown again if hydrochloric acid is then added.

2.5 Changing the conditions

An important reaction in the production of sulfuric acid is represented by the equation:

$$2SO_2(g) + O_2(g) \rightleftharpoons 2SO_3(g) + \text{heat}$$

1 What changes would occur to the amount of sulfur trioxide present at equilibrium in a reaction vessel for each of the following examples? Explain your answer.

a The partial pressure of oxygen gas is increased.

9780170412476

b The total pressure of the system is reduced by expanding the volume of the vessel.

c The temperature of the reaction mixture is increased.

2.6 Changes to pressure in gaseous reactions

The total pressure in a closed system containing nitrogen tetroxide (N_2O_4) and nitrogen dioxide (NO_2) at equilibrium is increased by pumping in some nitrogen gas. The equilibrium can be represented by the equation:

$$N_2O_4(g) \rightleftharpoons 2NO_2(g)$$

1 Suppose that sufficient nitrogen gas is pumped into the vessel for it to have a concentration of 4.00 M.

a Has the total pressure changed?

b Have the concentrations and the partial pressures of the reactant gases changed?

c Is this reaction at equilibrium?

EVALUATION

Aqueous iron thiocyanate ($FeSCN^{2+}$) has a distinctive red colour and can be formed by the reaction of yellow iron(III) ions (Fe^{3+}) with colourless thiocyanate ions (SCN^-). The reaction is:

$$Fe^{3+}(aq) + SCN^-(aq) \rightleftharpoons FeSCN^{2+}(aq)$$

1 In an experiment to investigate this reaction, various changes are made to a mixture at equilibrium containing the ions listed above. Complete Table 2.7.1 to predict the effect that the various changes listed would have on the reaction mixture.

TABLE 2.7.1 Colour and equilibrium changes in aqueous iron thiocyante

	COLOUR AT NEW EQUILIBRIUM COMPARED TO INITIAL EQUILIBRIUM		$[Fe^{3+}]$ AT NEW EQUILIBRIUM COMPARED WITH INITIAL EQUILIBRIUM	
CHANGES MADE	MORE RED	LESS RED	INCREASED	DECREASED
HPO_4^{2-} is added which forms colourless $FeHPO_4^+$				
Addition of a large volume of water				
$Fe^{3+}(aq)$ is added				
Hg^{2+} ions are added which forms a precipitate of $Hg(SCN)_2$				

2 The reaction is exothermic. Complete the graph below to show the changes that would occur to the concentrations of each species as the temperature increased.

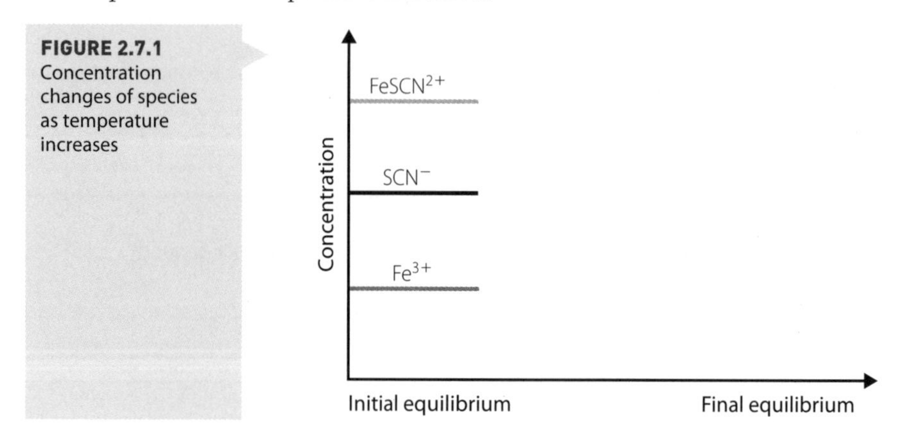

FIGURE 2.7.1 Concentration changes of species as temperature increases

3 Equilibrium constants

Summary

- The equilibrium constant, K_c, is a measure of the position of a system at equilibrium and is calculated as follows:

  For the reaction:

  $$aA + bB \rightleftharpoons cC + dD$$

  the equilibrium expression is:

  $$K_c = \frac{[C]^c [D]^d}{[A]^a [B]^b}$$

- The magnitude of K_c gives an indication of the extent of the reaction that is taking place.
  - If K_c is less than one, then there are more reactant molecules present than product molecules. The equilibrium is said to lie to the left.
  - If K_c is greater than one, there are more product molecules present than reactant molecules. The equilibrium is said to lie to the right.
  - If $K_c = 1$, the concentrations of product and reactant molecules present at equilibrium are approximately the same.
- The reaction quotient, Q, can be calculated in the same way as K_c, regardless of whether the system is at equilibrium or not. Therefore, if the value of K_c under some specific conditions is known, then comparing this with the measured value of Q, enables a determination of whether or not the system is at equilibrium.
- In order to calculate K_c, the concentrations of reactants and products at equilibrium must first be established. This can be done by using the reaction equation to determine the change in quantities that must have occurred from when the reaction was initiated to when equilibrium was established.

3.1 Important terms

1 Complete the crossword using the clues given.

Across	**Down**
1 Equilibrium ___________: the value of the reaction quotient when equilibrium has been established	**2** The only factor that can affect the equilibrium constant
3 A system where all the reactants and products are in the same phase	**4** Equilibrium ___________: concentrations of the products divided by the concentrations of the reactants, each multiplied together and raised to the powers of their coefficients in the reaction equations
5 Reaction ___________: the value of the equilibrium expression when calculated	
6 A system where there is a mixture of phases of reactants and products	
7 When the equilibrium position shifts so that the concentration of reactants increases, it moves to the ___________.	

 9780170412476

2 Write a short summary or create a mind map using all of the terms from the crossword.

3.2 Equilibrium constant expressions

Write the expressions that define the equilibrium constant for each of the following equilibrium reactions.

1 $CH_3COOH(aq) + H_2O(l) \rightleftharpoons CH_3COO^-(aq) + H_3O^+(aq)$

2 $Mg(OH)_2(s) \rightleftharpoons Mg^{2+}(aq) + 2OH^-(aq)$

3 $2SO_2(g) + O_2(g) \rightleftharpoons 2SO_3(g)$

4 $4Fe(s) + 3O_2(g) \rightleftharpoons 2Fe_2O_3(s)$

3.3 At equilibrium or not?

The data in Table 3.3.1 are relevant to the reaction represented by the equation, in reaction mixtures at 986°C:

$$H_2(g) + CO_2(g) \rightleftharpoons H_2O(g) + CO(g)$$

The initial concentrations of species in a reaction mixture, and the concentrations when the reaction has come to equilibrium, are indicated in the table.

TABLE 3.3.1 Initial concentration and equilibrium concentration for various species

	INITIAL CONCENTRATION (M)	EQUILIBRIUM CONCENTRATION (M)	CHANGE IN CONCENTRATION (M)	CONCENTRATION AFTER TIME HAD ELAPSED (M)
$H_2(g)$	1.0	0.44		
$CO_2(g)$	1.0	0.44		
$H_2O(g)$	0	0.56	−0.144	
$CO(g)$	0	0.56		

1 In another vessel at the same temperature, the initial concentrations of gases were $[H_2] = 0.200$ M, $[CO_2] = 0.400$ M, $[H_2O] = 0.600$ M, and $[CO] = 0.800$ M. After some time had elapsed, it was found that $[H_2O] = 0.456$ M. Is this reaction mixture at equilibrium? Explain your answer.

3.4 Calculating K_c

When one mole of $H_2(g)$ and 0.40 moles of $N_2(g)$ are placed in a 5.0 L container and allowed to reach an equilibrium at a certain temperature, it is found that 0.078 moles of NH_3 is present at equilibrium. The reaction is:

$$2NH_3(g) \rightleftharpoons 3H_2(g) + N_2(g)$$

1 Use the equation to determine how many moles of nitrogen and hydrogen are present at equilibrium.

9780170412476

2 Calculate the concentration of each of the three species at equilibrium.

3 Write the expression for K_c and calculate its value.

3.5 Using equilibrium constants to predict quantities

The equilibrium constant for the reaction represented below at 448°C is 50.

$$H_2(g) + I_2(g) \rightleftharpoons 2HI(g)$$

1 If 1 mole of H_2 is mixed with 1 mole of I_2 in a 0.50 L container and allowed to react at 448°C, when equilibrium is reached, 0.03 moles of I_2 is remaining. How many moles of HI will have been produced?

Nitrogen dioxide $NO_2(g)$ reacts to form dinitrogen tetroxide, $N_2O_4(g)$, as shown by the equation:

$$2NO_2(g) \rightleftharpoons N_2O_4(g)$$

In an experiment, 1.42 moles of NO_2 is placed in an empty 2 L vessel at 200°C. When the system reaches equilibrium, 0.30 moles of N_2O_4 is present in the reaction vessel.

1 Calculate the equilibrium constant for this reaction.

2 At 100°C, the numerical value of the equilibrium constant for this reaction is 12.05. Is this reaction exothermic or endothermic? Explain your answer.

4 Properties of acids and bases

LEARNING

Summary

- When an acid dissociates in solution it produces the hydronium ion, $H_3O^+(aq)$.
- The strength of an acid or a base depends on the degree to which it dissociates in solution.
- Monoprotic acids dissociate in one step, producing one hydronium ion per molecule of acid.
- Polyprotic acids are acids that can dissociate via more than one step, thereby producing two or more hydronium ions.
- Polyprotic acids can produce a number of anions during dissociation that can then combine with a cation to produce hydrogen salts or acid salts.
- A strong acid is one that dissociates completely in water.
- Strong acids include hydrochloric acid, HCl, nitric acid, HNO_3, and sulfuric acid, H_2SO_4.
- A weak acid is one that only partially dissociates in water.
- Weak acids include ethanoic acid, CH_3COOH, carbonic acid, H_2CO_3, hydrofluoric acid, HF, and citric acid, $C_6H_8O_7$.
- A base is a substance that produces hydroxide, OH^-, ions in solution.
- A strong base is one that dissociates completely in water.
- Strong bases include the hydroxides and oxides of groups 1 and 2.
- Weak bases include ammonia, NH_3, and methylamine, CH_3NH_2.
- Concentrated acids and bases contain relatively large number of acid or base molecules per unit volume; dilute acids and bases contain few acid or base molecules per unit volume.
- The strength of an acid or base can be measured by its electrical conductivity. A strong acid or base will dissociate completely, producing large numbers of ions, giving a high conductivity. A weak acid or base will produce relatively few ions, giving a low conductivity.

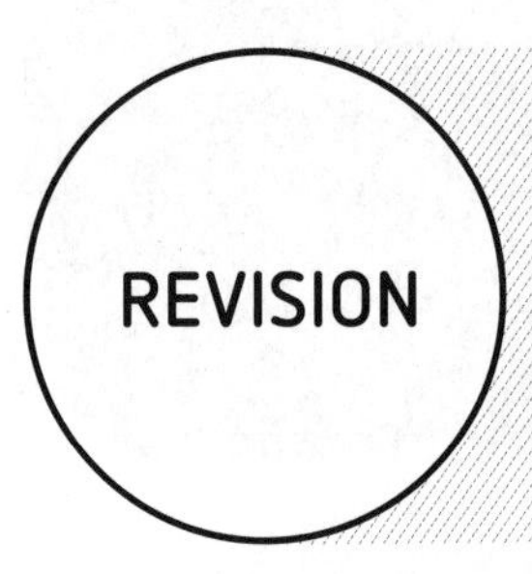

4.1 Concept map

The following shows a concept map for properties of acids and bases. Some of the concepts have been provided, the rest are given in the concept list. Use the list to complete the concept map.

Concept list

carbonic	ethanoic	high	hydrochloric
low	methylamine	nitric	OH^-
potassium	sodium	strong	strong
sulfuric	weak	weak	

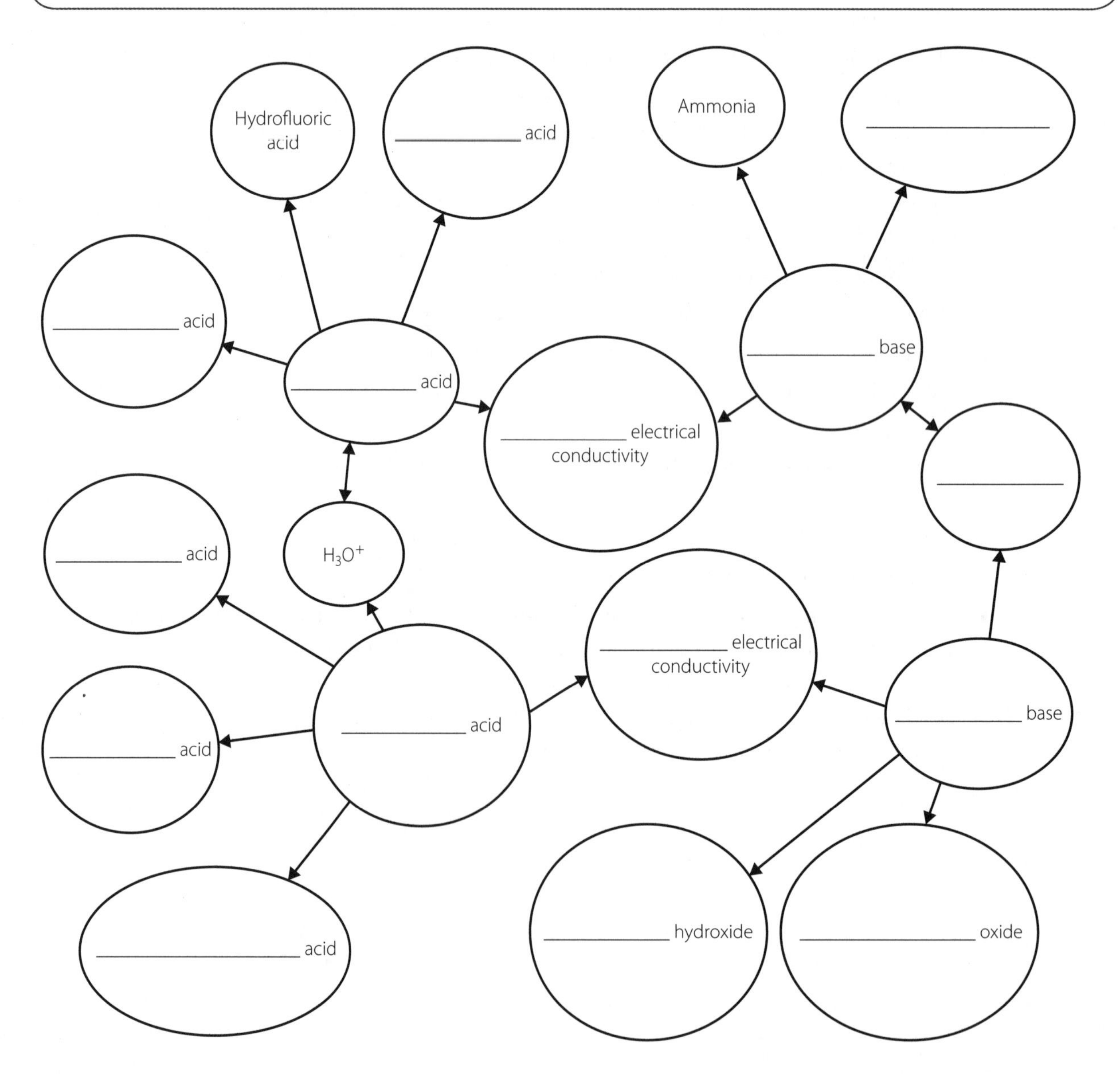

4.2 Acids and bases

Previously it was shown that the Arrhenius definition of an acid was a substance that produces H^+ ions in solution.

BUT

An H^+ ion is simply a proton that is a very small, highly concentrated positive charge. In solution, the proton immediately attaches itself to a water molecule:

$$H^+(aq) + H_2O(l) \rightarrow H_3O^+(aq)$$

Hydronium ion

Therefore, the Arrhenius definition of an acid should be modified to:

An acid is a substance that produces H_3O^+ ions in solution.

POLYPROTIC ACIDS

- *Monoprotic acids* dissociate in one step:

$$HCl(aq) + H_2O(l) \rightarrow H_3O^+(aq) + Cl^-(aq)$$

- *Diprotic acids* dissociate in two steps:

$$H_2SO_4(l) + H_2O(l) \rightarrow HSO_4^-(aq) + H_3O^+(aq)$$

Hydrogen sulfate ion

$$HSO_4^-(aq) + H_2O(l) \rightarrow SO_4^{2-}(aq) + H_3O^+(aq)$$

Sulfate ion

- *Triprotic acids* dissociate in three steps:

$$H_3PO_4(s) + H_2O(l) \rightarrow H_2PO_4^-(aq) + H_3O^+(aq)$$

Dihydrogen phosphate ion

$$H_2PO_4^-(aq) + H_2O(l) \rightarrow HPO_4^{2-}(aq) + H_3O^+(aq)$$

Hydrogen phosphate ion

$$HPO_4^{2-}(aq) + H_2O(l) \rightarrow PO_4^{3-}(aq) + H_3O^+(aq)$$

Phosphate ion

1 Write definitions for the following terms.

a acid ______________________________

b base ______________________________

c hydronium ion ______________________________

d monoprotic acid ______________________________

e diprotic acid ______________________________

f triprotic acid ______________________________

g weak acid ______________________________

h weak base ______________________________

2 Fill in the blanks in the following sentences.

a A strong __________ is one that dissociates completely to produce OH^- ions in solution. Strong bases include__________ __________ and __________ __________.

b A weak acid such as__________ __________ only __________ dissociates to produce __________ in solution.

c A diprotic acid dissociates to produce __________ __________ ions per molecule of acid.

4.3 Acid and base strength

The strength of an acid is based on the number of hydronium ions it puts into solution. This is based on the degree to which an acid dissociates.

A strong acid such as nitric acid dissociates completely:

$$HNO_3(aq) + H_2O(l) \rightarrow H_3O^+(aq) + NO_3^-(aq)$$
$$0\% \qquad\qquad 100\%$$

A weak acid such as ethanoic acid only partially dissociates:

$$CH_3COOH(aq) + H_2O(l) \rightleftharpoons H_3O^+(aq) + CH_3COO^-(aq)$$
$$\approx 99\% \qquad\qquad \approx 1\%$$

Sodium hydroxide (NaOH) is a strong base and so it dissociates fully:

$$NaOH(s) + H_2O(l) \rightleftharpoons Na^+(aq) + H_2O(l) + OH^-(aq)$$
$$0\% \qquad\qquad 100\%$$

Methylamine (CH_3NH_2) is a weak base and only partially dissociates:

$$CH_3NH_2(l) + H_2O(l) \rightleftharpoons CH_3NH_3^+(aq) + OH^-(aq)$$
$$\approx 99\% \qquad\qquad \approx 1\%$$

1 Fill in the blanks in the following sentences.

a __________ __________ dissociates completely to produce __________ ions and sulfate ions.

b __________ __________ partially dissociates to produce __________ ions and ethanoate ions.

c Phosphoric acid is a __________ acid because it dissociates in _____ stages.

d Carbonic acid dissociates in two stages to give hydronium ions and __________ __________ ions in the first stage, and hydronium ions and __________ ions in the second stage.

4.4 Concentration versus strength

1 Fill in the missing words.

a The __________ of an acid or a base refers to the degree of dissociation that occurs and, therefore, the number of hydronium or hydroxide ions produced in solution.

b The __________ of an acid or a base describes how many molecules of acid or base are present per unit volume of solution.

c Acids and bases can be strong or__________.

d Acids and bases can be concentrated or __________.

A strong acid could be dilute yet still produce more hydronium ions in solution than a concentrated weak acid. A useful way to differentiate between strong and weak acids is to measure their electrical conductivity.

Electrical conductivity is a measure of how many ions are in solution. If equal concentrations of acids and bases are used, strong acids and bases will produce more ions in solution that will lead to greater electrical conductivity.

1 Chemicals in which almost all the molecules will lose a hydrogen ion in solution are called:

A strong bases.

B weak acids.

C strong acids.

D weak bases.

2 Which of the following is NOT a weak acid?

A HF

B HCl

C CH_3COOH

D H_2CO_3

3 The name given to acids that are able to donate more than one proton is:

A diprotic.

B triprotic.

C polyprotic.

D strong acid.

4 Which solution would have the greatest electrical conductivity?

A Diprotic

B Triprotic

C Polyprotic

D Strong acid

5 Write equations to show the three stages in the dissociation of phosphoric acid in water. At each stage, give the name of the anion formed.

____

____

____

____

____

____

6 Explain why a 0.1 M solution of nitric acid dissociates to produce more H_3O^+ ions in solution than a 5 M solution of carbonic acid.

____

____

____

____

7 An experiment is carried out in which a solution of sodium hydroxide is added to 20 mL of 1 M HCl. The conductivity of the solution is measured throughout. Figure 4.5.1 shows a graph of conductivity of the solution against volume of base added.

FIGURE 4.5.1 Conductivity of a solution against volume of base added

a As the base is added to the acid, the conductivity of the solution changes. Explain the conductivity of the solution at each of the points marked 1–5 on the graph.

1 ____

2 ____

3 ____

4 ____

5 ____

b From the shape of the graph, estimate the concentration of the sodium hydroxide solution. Explain your answer.

____

____

____

____

c On the graph above, draw sketches of the graphs you would expect if the following reactions were to take place.

i 1 M of HCl with 1 M of NH_3

ii 1 M of CH_3COOH with 1 M of NaOH

iii 1 M of CH_3COOH with 1 M of NH_3

5 pH scale

Summary

- In pure water, a few water molecules react with each other to produce hydronium ions and hydroxide ions. This reaction is called the *self-ionisation of water*.
- The equation for the self-ionisation of water is given by:

$$H_2O(l) + H_2O(l) \rightleftharpoons H_3O^+(aq) + OH^-(aq)$$

- The equilibrium expression for the self-ionisation of water is given by:

$$K_w = [H_3O^+]\,[OH^-]$$

- At room temperature, 298 K, $K_w = 1 \times 10^{-14}\,M^2$.
- Therefore, in pure water: $[H_3O^+] = [OH^-] = 1 \times 10^{-7}\,M$.
- Using this relationship, the concentration of H_3O^+ and OH^- can be determined for acidic and basic solutions.
- When $[H_3O^+] > [OH^-]$, the solution is considered acidic. At 298 K, $[H_3O^+] > 10^{-7}\,M$.
- When $[H_3O^+] < [OH^-]$, the solution is considered basic. At 298 K, $[H_3O^+] < 10^{-7}\,M$.
- The pH of an acid solution is given by:

$$pH = -\log_{10}[H_3O^+]$$

- The pH of a basic solution can be calculated using the self-ionisation of water and the equation given above.
- The pOH of a basic solution is given by:

$$pOH = -\log_{10}[OH^-]$$

- The relationship between pH and pOH is expressed by the equation:

$$pH + pOH = 14$$

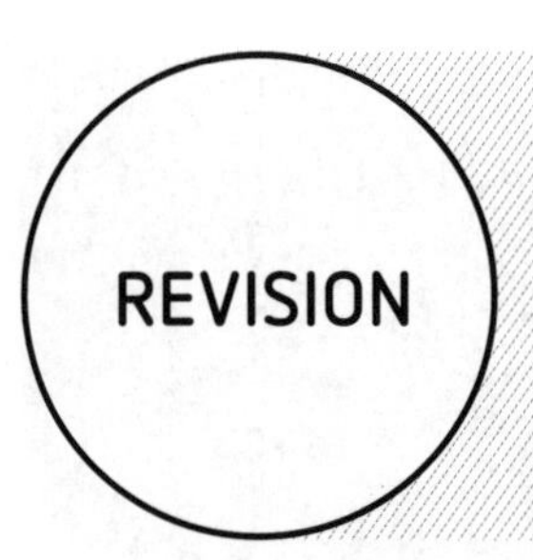

5.1 Concept map

The following is a concept map for pH scale. Some of the concepts have been provided, the rest are given in the concept list. Use the list to complete the concept map.

Concept list

1	1	7	14	14
14	$[H_3O^+]$	$[H_3O^+]$	$[H_3O^+]$	$[OH^-]$
$[OH^-]$	$[OH^-]$	pH	pOH	$[H_3O^+]$
				$[OH^-]$

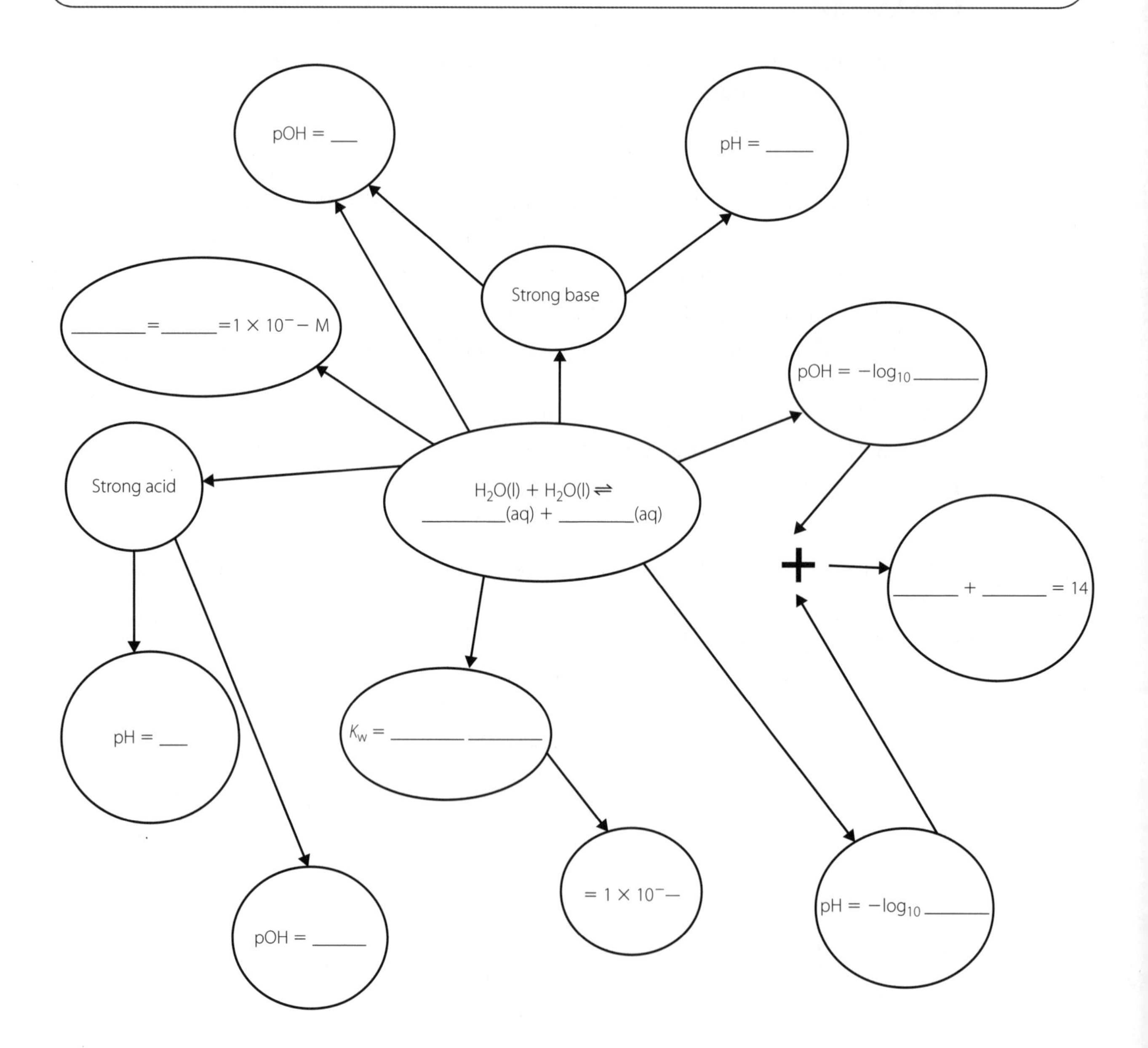

9780170412476

5.2 Self-ionisation of water

In pure water, a small number of water molecules react together according to the equation:

$$H_2O(l) + H_2O(l) \rightleftharpoons H_3O^+(aq) + OH^-(aq)$$

This is an equilibrium reaction and so has an equilibrium constant called the self-ionisation constant, K_w.

According to the rules for writing equilibrium expressions, the $H_2O(l)$ does not appear in the expression and so the equilibrium expression for the self-ionisation of water becomes:

$$K_w = [H_3O^+]\,[OH^-]$$

At room temperature, 298 K, the value of K_w is 1×10^{-14} M^2.

This becomes very useful when the concentrations of H_3O^+ *and* OH^- need to be calculated.

WORKED EXAMPLE 1

Calculate the concentration of H_3O^+ and OH^- ions in a 2.95 M solution of nitric acid.

ANSWER

a Write the dissociation equation.

$$HNO_3(aq) + H_2O(l) \rightarrow H_3O^+(aq) + NO_3^-(aq)$$

b Determine the H_3O^+ concentration from the nitric acid concentration.

Nitric acid is a strong, monoprotic acid; therefore, the $[H_3O^+]$ is the same as the nitric acid concentration.

$$[H_3O^+] = 2.95\text{ M}$$

c Determine the $[OH^-]$ using $K_w = [H_3O^+]\,[OH^-] = 1 \times 10^{-14}$

If

$$[H_3O^+]\,[OH^-] = 1 \times 10^{-14}$$

then

$$[OH^-] = \frac{1 \times 10^{-14}}{2.95}$$

$$= 3.39 \times 10^{-15}\text{ M}$$

WORKED EXAMPLE 2

Calculate the concentration of H_3O^+ and OH^- ions in a 8.34×10^{-2} M solution of strontium hydroxide, $Sr(OH)_2$.

ANSWER

a Write the dissociation equation.

$$Sr(OH)_2(aq) \rightarrow Sr^{2+}(aq) + 2OH^-(aq)$$

b Determine the OH^- concentration from the strontium hydroxide concentration.

Strontium hydroxide, being a group 2 hydroxide, is strong. For every one mole of $Sr(OH)_2$, two moles of OH^- ions is produced.

Therefore:

$$[OH^-] = 2 \times 8.34 \times 10^{-2}$$

$$= 0.17\text{ M}$$

c Determine the $[H_3O^+]$ using $K_w = [H_3O^+]\,[OH^-] = 1 \times 10^{-14}$.

$$[OH^-] = \frac{1 \times 10^{-14}}{0.17}$$

$$= 5.88 \times 10^{-14}\text{ M}$$

QUESTIONS

1 Calculate the $[H_3O^+]$ and $[OH^-]$ in the following solutions.

a 0.74 M HCl

b 1.39 M NaOH

c 2.45×10^{-4} M H_2SO_4

d 1.15×10^{-4} M $Ba(OH)_2$

5.3 pH scale

The pH scale is a measure of the concentration of hydronium ions in solution. In acidic and alkaline solutions, the concentration of hydronium ions can vary greatly, from 10 M to 10^{-15} M. pH is a simple and convenient way of expressing these, sometimes unwieldly, numbers.

The pH of a solution is defined as the negative logarithm (to the base 10) of the hydronium ion concentration:

$$pH = -\log_{10} [H_3O^+]$$

For a strong monoprotic acid such as HCl the pH calculation is straightforward:

If

$$[HCl] = 0.16\,M, [H_3O^+] = 0.16\,M$$

$$pH = -\log_{10} 0.16$$

$$= 0.79$$

Calculating the pH of alkaline solutions requires an extra step and uses the equation for the self-ionisation of water discussed in section 5.2.

To calculate the pH of a 0.14 M NaOH solution:

If

$$[NaOH] = 0.14\,M, [OH^-] = 0.14\,M$$

Substituting this into

$$[H_3O^+]\,[OH^-] = 1 \times 10^{-14}$$

gives

$$[H_3O^+] \times 0.14 = 1 \times 10^{-14}$$

Therefore,

$$[H_3O^+] = \frac{1 \times 10^{-14}}{0.14}$$

$$= 7.14 \times 10^{-14}\,M$$

The pH can now be calculated:

$$pH = -\log_{10} 7.14 \times 10^{-14}$$

$$= 13.15$$

pOH

In the equation $pH = -\log_{10} [H_3O^+]$, the letter p is a form of standard notation meaning that $-\log_{10}$ of whatever comes after it.

So, the pOH of an alkaline solution is given by:

$$pOH = -\log_{10} [OH^-]$$

The pOH of a 0.28 M KOH solution can be calculated:

$$pOH = -\log_{10} 0.28$$

$$= 0.55$$

A very useful relationship exists between pH and pOH, and is given by:

$$pH + pOH = 14$$

WORKED EXAMPLE 1

Calculate the pH of a 0.086 M NaOH solution.

ANSWER

a Calculate pOH.

$$pOH = -\log_{10} 0.086$$

$$= 1.07$$

b Calculate pH.

$$pH = 14 - 1.07$$

$$= 12.93$$

QUESTIONS

1 Calculate the pH and pOH of the following solutions.

a $0.0043\,M\ HNO_3$

b $0.98\,M\ Sr(OH)_2$

c $1.76 \times 10^{-4}\,M\ H_2SO_4$

d $3.25\,M\ KOH$

2 Calculate the $[H_3O^+]$ in the following solutions.

a NaOH pH 11.25

b $Ba(OH)_2$ pH 13.10

3 Calculate the $[OH^-]$ in the following solutions.

a HNO_3 pH 3.4

b H_2SO_4 pH 0.24

4 Figure 5.3.1 shows the relationship between the pH and pOH scales. Fill in the blank $[H_3O^+]$ and $[OH^-]$ boxes.

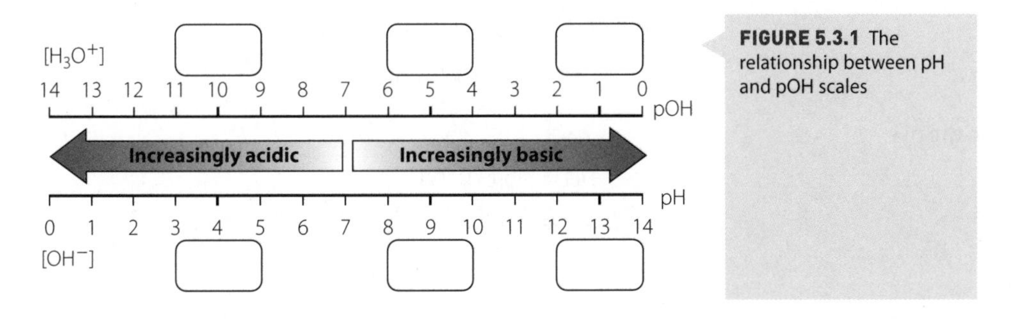

FIGURE 5.3.1 The relationship between pH and pOH scales

RESULTING PH OF ACID–BASE MIXTURES

When acids and bases are mixed, neutralisation takes place according to the equation:

$$H_3O^+(aq) + OH^-(aq) \rightarrow 2H_2O(l)$$

However, if one of the reactants is in excess, the resulting solution will have a pH value dependent on whether the acid or the base was in excess.

WORKED EXAMPLE 2

Calculate the pH of a solution in which 25.8 mL of 0.982 M nitric acid are mixed with 29.6 mL of 0.738 M NaOH.

ANSWER

a Write the neutralisation equation to determine the mole ratio.

$$HNO_3(aq) + NaOH \rightarrow NaNO_3(aq) + H_2O(l)$$

1 mol HNO_3 reacts with 1 mol NaOH

b Draw a diagram. It can be helpful here to draw a diagram to represent the reaction.

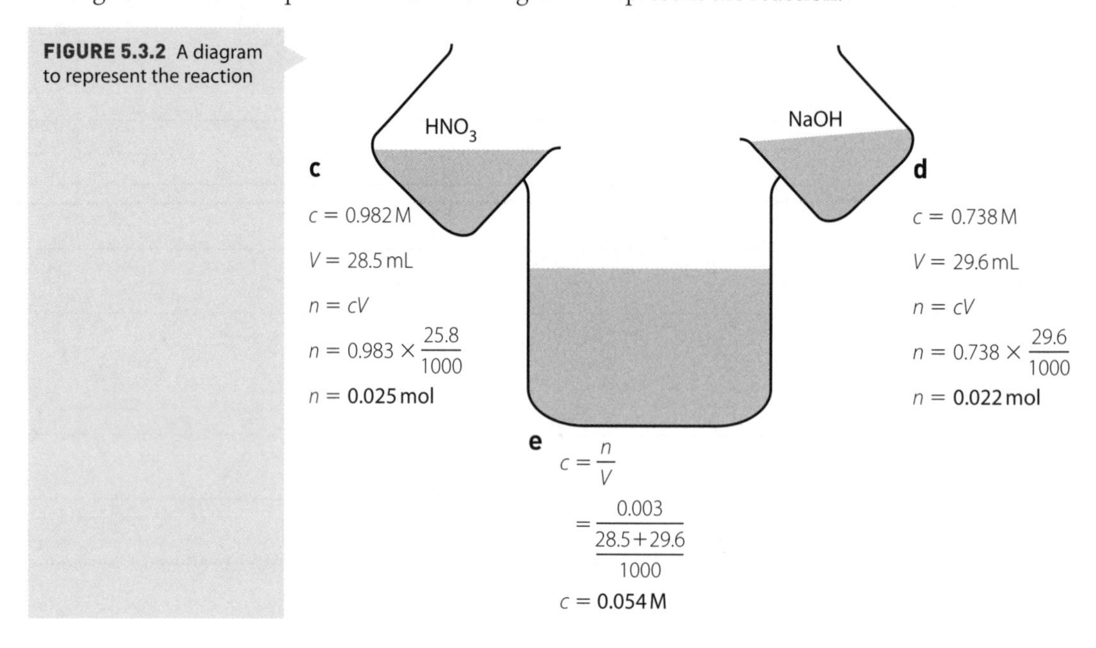

FIGURE 5.3.2 A diagram to represent the reaction

f Determine the excess.

With 0.025 M, HNO_3 is in excess. Most of this will be neutralised by the 0.022 mol of the NaOH.

$$\begin{aligned} HNO_3(\text{excess}) &= 0.025 - 0.022 \\ &= 0.003\ \text{mol} \end{aligned}$$

g Determine the pH.

$$\begin{aligned} pH &= -\log_{10} 0.054 \\ &= 1.27 \end{aligned}$$

QUESTION

5 Calculate the pH of the mixture produced by mixing 128.3 mL of 1.130 M HCl and 329.7 mL of 0.997 M NaOH.

9780170412476

1 In pure water:

A the pH is always 7.

B there are only water molecules.

C the concentration of hydrogen ions is the same as the concentration of hydroxide ions.

D there are no ions present.

2 In a solution with a concentration of H_3O^+ ions of 0.1, the concentration of hydroxide ions is:

A 0.1 M.

B 10^{-12} M.

C 10^{-13} M.

D 1 M.

3 The concentration of H_3O^+ ions in a 1.5 mol L^{-1} solution of sodium hydroxide is:

A 1×10^{-14} M.

B 6.67×10^{15} M.

C 12.5 M.

D 1.5×10^{14} M.

4 What is the pOH of a 0.086 M solution of H_2SO_4?

A 12.94

B 0.76

C 0.17

D 13.24

5 Solution X has a pH of 3. How many more H_3O^+ ions does it have compared to solution Y, which has a pOH of 8? (Hint: refer back to Fig. 5.3.1 to help you answer this question.)

6 At 40°C, the value for K_w is 2.9×10^{-14}. What is the pH of pure water at this temperature? Remember that in pure water, the concentration of H^+ equals the concentration of OH^-

7 It is *not* strictly true to say that the pH scale ranges from 1 to 14. Explain what this statement means with the aid of example calculations.

8 Calculate the pH of the solution resulting from mixing 53.15 mL of 1.04 M H_2SO_4 with 78.34 mL of 1.98 M KOH.

9780170412476

9 Fill in the blanks in the table below.

$[H_3O^+]$	$[OH^-]$	pH	pOH	ACIDIC/BASIC
			4.5	
3.91×10^{-9}				
	4.7×10^{-11}			
		0.8		
			1.3	
5.3×10^{-7}				

6 Brønsted–Lowry model

Summary

- The Danish chemist Johannes Brønsted and the English chemist Thomas Lowry identified the need to refine the definition of an acid and a base.
- According to the Brønsted–Lowry model:
  - an acid is a proton donor
  - a base is a proton acceptor.
- A proton is simply a hydrogen ion, H^+.
- A conjugate base is produced when an acid donates its proton to a base.
- A conjugate acid is produced when a base accepts a proton from an acid.
- In an acid–base reaction the acid and the conjugate base it produces is called a conjugate pair. The same goes for a base and the conjugate acid it produces.
- An amphiprotic substance is one that has the ability to donate or accept a proton.
- A buffer solution is one that can maintain a pH within a certain range despite the addition of an acid or a base.
- An acidic buffer solution has a pH less than 7 and is composed of a weak acid and one of its salts.
- A basic buffer solution has a pH greater than 7 and is composed of a weak base and one of its salts.

9780170412476

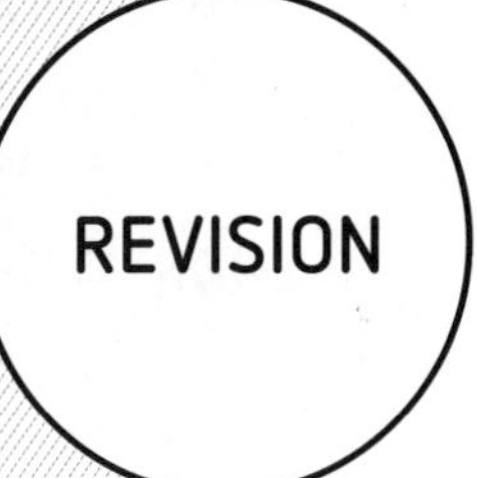

6.1 Concept map

The following shows a concept map for Brønsted–Lowry model. Some of the concepts have been provided, the rest are given in the concept list. Use the list to complete the concept map.

Concept list

acceptor	acid	acid	base
base	donor	weak	weak

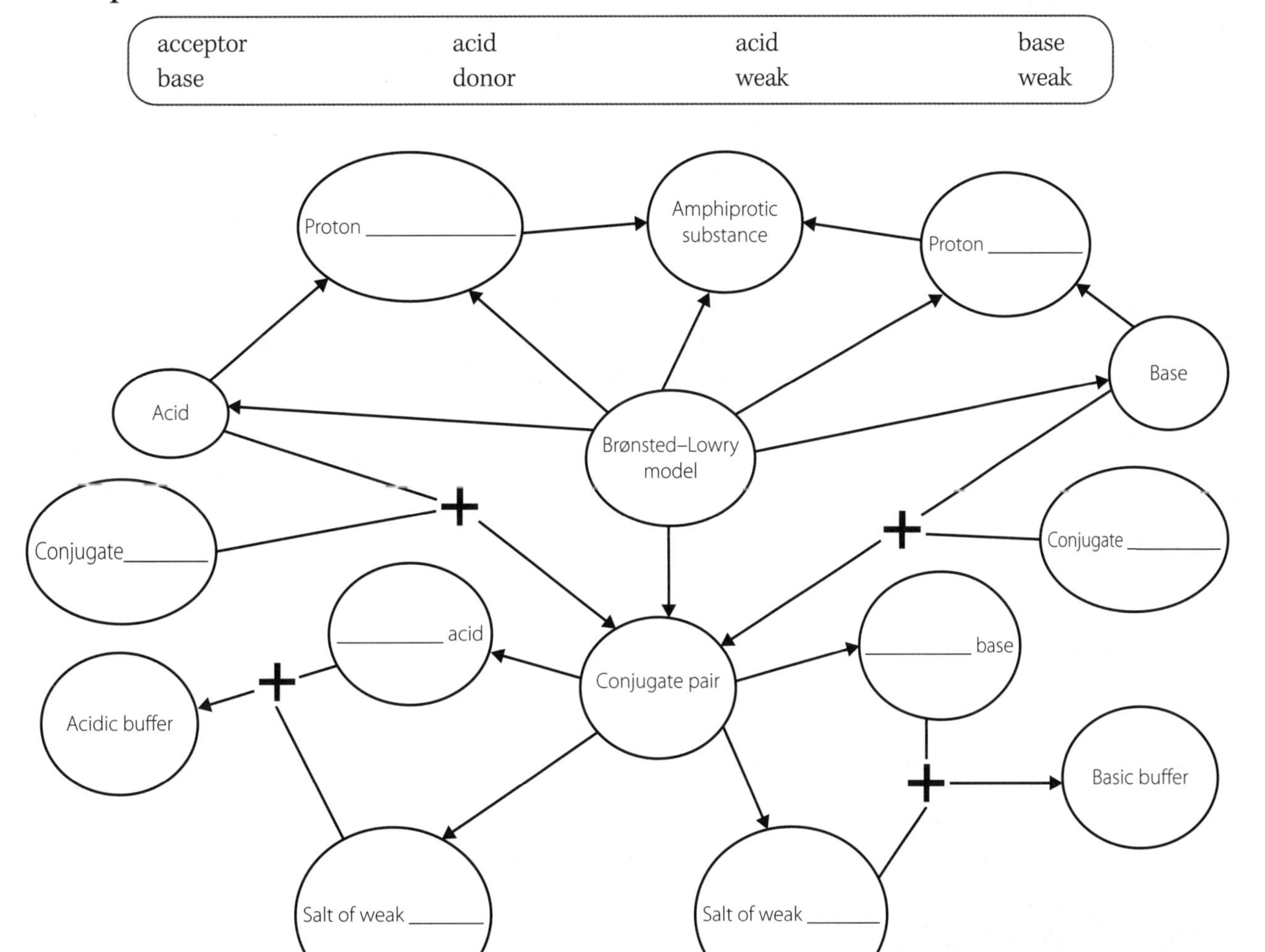

6.2 Brønsted–Lowry model

The Brønsted–Lowry model states that:

- an acid is a proton donor
- a base is a proton acceptor.

When an acid has donated its proton, the species left behind is called a conjugate base.
When a base accepts a proton, the species produced is called a conjugate acid.
This is most easily seen when reacting the acid or the base with water.

BRØNSTED–LOWRY ACID

FIGURE 6.2.1 Equation showing hydrochloric acid acting as a Brønsted–Lowry acid

$$\underset{\text{Acid}}{HCl(aq)} + \underset{\text{Base}}{H_2O(l)} \rightarrow \underset{\text{Conjugate base}}{Cl^-(aq)} + \underset{\text{Conjugate acid}}{H_3O^+(aq)}$$

$-H^+$ (HCl → Cl^-); $+H^+$ (H_2O → H_3O^+)

In this case:

- HCl/Cl^- are a conjugate pair
- H_2O/H_3O^+ are a conjugate pair.

BRØNSTED–LOWRY BASE

FIGURE 6.2.2 Equation showing ammonia acting as a Brønsted–Lowry base

$$\underset{\text{Base}}{NH_3(aq)} + \underset{\text{Acid}}{H_2O(l)} \rightarrow \underset{\text{Conjugate acid}}{NH_4^+(aq)} + \underset{\text{Conjugate base}}{OH^-(aq)}$$

$+H^+$ (NH_3 → NH_4^+); $-H^+$ (H_2O → OH^-)

In this case:

- NH_3/NH_4^+ are a conjugate pair
- H_2O/OH^- are a conjugate pair.

1 In the equations below, identify the acid (A), base (B), conjugate acid (CA) and conjugate base (CB).

a $HCl(aq) + NH_3(aq) \rightarrow NH_4^+(aq) + Cl^-(aq)$

______ ______ ______ ______

b $H_2PO_4^-(aq) + HI(aq) \rightarrow H_3PO_4(aq) + I^-(aq)$

______ ______ ______ ______

c $H_2S(aq) + NH_2^-(aq) \rightarrow HS^-(aq) + NH_3(aq)$

______ ______ ______ ______

d $CH_3COOH(aq) + CH_3CH_2NH_2(aq) \rightarrow CH_3COO^-(aq) + CH_3CH_2NH_3^+(aq)$

______ ______ ______ ______

6.3 Amphiprotism

An amphiprotic substance is one that can act as either a Brønsted–Lowry acid or a Brønsted–Lowry base.
Consider the self-ionisation of water:

$$\underset{\text{Acid}}{H_2O(l)} + \underset{\text{Base}}{H_2O(l)} \rightleftharpoons \underset{\text{Conjugate acid}}{H_3O^+(aq)} + \underset{\text{Conjugate base}}{OH^-(aq)}$$

1 For the following reactions, fill in the blanks to show the reacting species acting as an acid or a base:

a Acid

$HCO_3^-(aq) + H_2O(aq) \rightarrow$ __________ + __________

Base

$HCO_3^-(aq) + H_2O(aq) \rightarrow$ __________ + __________

9780170412476

b Acid

$$H_2PO_4^-(aq) + H_2O(aq) \rightarrow \_\_\_\_\_\_\_\_ + \_\_\_\_\_\_\_\_$$

Base

$$H_2PO_4^-(aq) + H_2O(aq) \rightarrow \_\_\_\_\_\_\_\_ + \_\_\_\_\_\_\_\_$$

c Acid

$$H_2NCH_2COOH(aq) + H_2O(aq) \rightarrow \_\_\_\_\_\_\_\_ + \_\_\_\_\_\_\_\_$$

Base

$$H_2NCH_2COOH(aq) + H_2O(aq) \rightarrow \_\_\_\_\_\_\_\_ + \_\_\_\_\_\_\_\_$$

6.4 Buffers

A buffer solution is able to maintain a certain pH range when small amounts of acid or base are added to it.

Buffer solutions can be classified into two broad types.

1 *Acidic buffer solutions*

These generally consist of a weak acid and the salt of its conjugate base.

For example, the ethanoic acid/sodium ethanoate buffer, CH_3COOH/CH_3COONa:

$$CH_3COOH(aq) + H_2O(aq) \rightleftharpoons CH_3COO^-(aq) + H_3O^+(aq)$$

- Adding a small amount of acid (H_3O^+):
  This disturbs the equilibrium.
  According to Le Chatelier's principle, the system will shift in the direction that will minimise the change. In other words, it will move in the direction that uses up the extra H_3O^+. It achieves this by shifting in the reverse direction; in other words, moving to the left.
  The CH_3COO^- ions react with the H_3O^+ until equilibrium is restored.
- Adding a small amount of base (OH^-):
  The system shifts to minimise the effect of the extra OH^- ions. More CH_3COOH molecules dissociate, producing more H_3O^+ ions that neutralise the extra OH^- ions until equilibrium is restored.

2 *Basic buffer solutions*

These consist of a weak base and the salt of its conjugate acid.

For example, the ammonia/ammonium chloride buffer, NH_3/NH_4Cl:

$$NH_3(aq) + H_2O(l) \rightleftharpoons NH_4^+(aq) + OH^-(aq)$$

- Adding a small amount of acid (H_3O^+):
  After an initial lowering of pH due to the extra H_3O^+ ions, the system shifts to the right as more NH_3 molecules react, producing more OH^- ions that neutralise the H_3O^+ thereby restoring equilibrium.

QUESTION

1 Using the information given above, suggest and describe a reaction shift that would restore equilibrium when a small amount of base (OH^-) is added to a basic buffer solution.

____

____

____

____

1 According to Brønsted–Lowry, which of the following statements is correct?

A An acid is a proton acceptor.

B A base is a proton donor.

C A base is only able to react with acids.

D An acid is a proton donor.

2 A buffer:

A works by shifting the position of equilibrium resulting in the concentration of hydrogen ions remaining approximately constant.

B is a solution that is able to resist a change in pH when an acid or a base is added.

C is a solution made of a weak acid and its conjugate base.

D all of the above.

3 Which one of the following is the conjugate base of the $H_2PO_4^-$ ion?

A HPO_4^{2-}

B H_3PO_4

C H_3O^+

D OH^-

4 If a small amount of base is added to an acidic buffer solution:

A the system shifts to the left to reduce the amount of base.

B the system shifts to the right, producing more H_3O^+ to neutralise the base.

C the system shifts to the right, producing more conjugate base, masking the OH^-.

D the system shifts to the left because the conjugate base removes the OH^-.

5 Write dissociation in water equations for the following substances. Identify the acid, base, conjugate acid and conjugate base in each equation.

a NH_3

____

____

b H_2SO_4 (both stages)

____

____

9780170412476

c HCO_3^- (acting as acid)

d H_3BO_3 (all three stages)

6 An experiment was carried out with the methylamine/methylamine ammonium chloride buffer system:

$$CH_3NH_2(aq) + H_2O(l) \rightarrow CH_3NH_3^+(aq) + OH^-(aq)$$

A number of changes were made to the system at various times and the concentrations of $CH_3NH_2(aq)$, $CH_3NH_3^+(aq)$ and $OH^-(aq)$ were measured. Figure 6.5.1 shows the graph of the results.

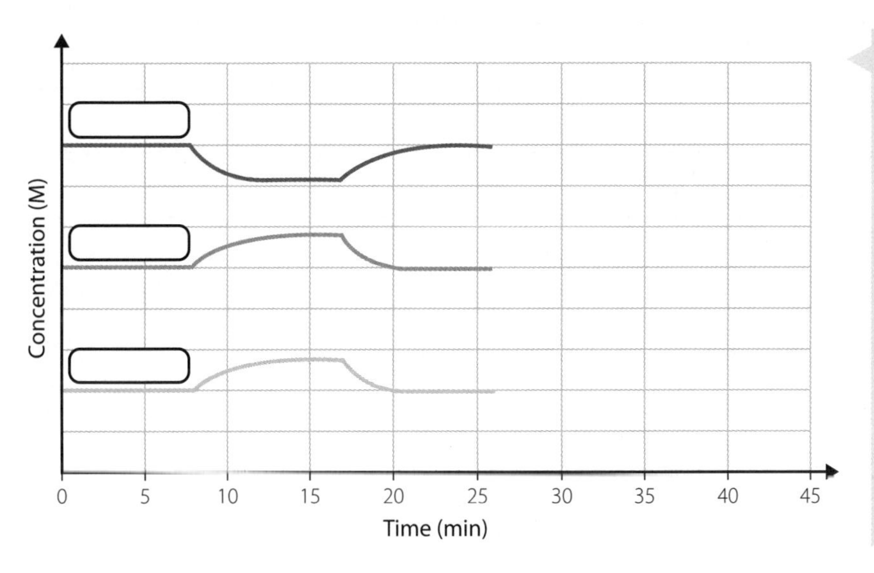

FIGURE 6.5.1 Graph of concentration against time for a buffer system

a What change occurred at time, $t = 8$ min? Explain your answer.

b What change occurred at time, $t = 17$ min? Explain your answer.

c Fill in the empty boxes on the graph.

d At $t = 26$ min a large volume of water was added to the system. On the graph, draw what happened to the concentration of each species in the system up to $t = 40$ min. Explain your reasoning in each case.

7 Dissociation constants

Summary

- In weak acids, only a small number of molecules dissociate, meaning that there is an equilibrium between the acid and its conjugate base. For example, HF is a weak acid:

$$HF(aq) + H_2O(l) \rightleftharpoons F^-(aq) + H_3O^+(aq)$$

The equilibrium expression for the dissociation of a weak acid is represented by:

$$K_a = \frac{[H_3O^+][F^-]}{[HF]}$$

where K_a is called the acid dissociation constant.

- K_a values are a measure of the strength of acids.
  - Strong acids have very large K_a values (greater than 1).
  - Weak acids have small K_a values (typically between around 10^{-1} and 10^{-14}).
- In weak bases, only a small fraction of the base molecules dissociate, giving rise to an equilibrium between the base and its conjugate acid. For example, NH_3 is a weak base:

$$NH_3(aq) + H_2O(l) \rightleftharpoons NH_4^+(aq) + OH^-(aq)$$

- The equilibrium expression for this reaction is given by:

$$K_b = \frac{[NH_4^+][OH^-]}{[NH_3]}$$

where K_b is called the base dissociation constant.

- Weak bases have small K_b values.
- K_a values can be determined from pH values and vice versa.
- Percentage ionisation of an acid can be determined from pH values and vice versa.
- An important relationship exists between K_a and K_b, and is given by:

$$K_w = K_a \times K_b$$

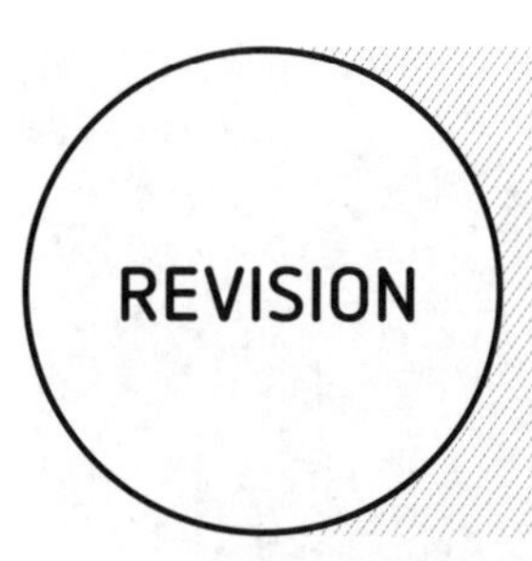

7.1 Concept map

The following shows a concept map for dissociation constants. Some of the concepts have been provided, the rest are given in the concept list. Use the list to complete the concept map.

Concept list

acid	weak	B	weak
H_3O	H_3O^+	HA	K_w
OH^-	pH	HB	

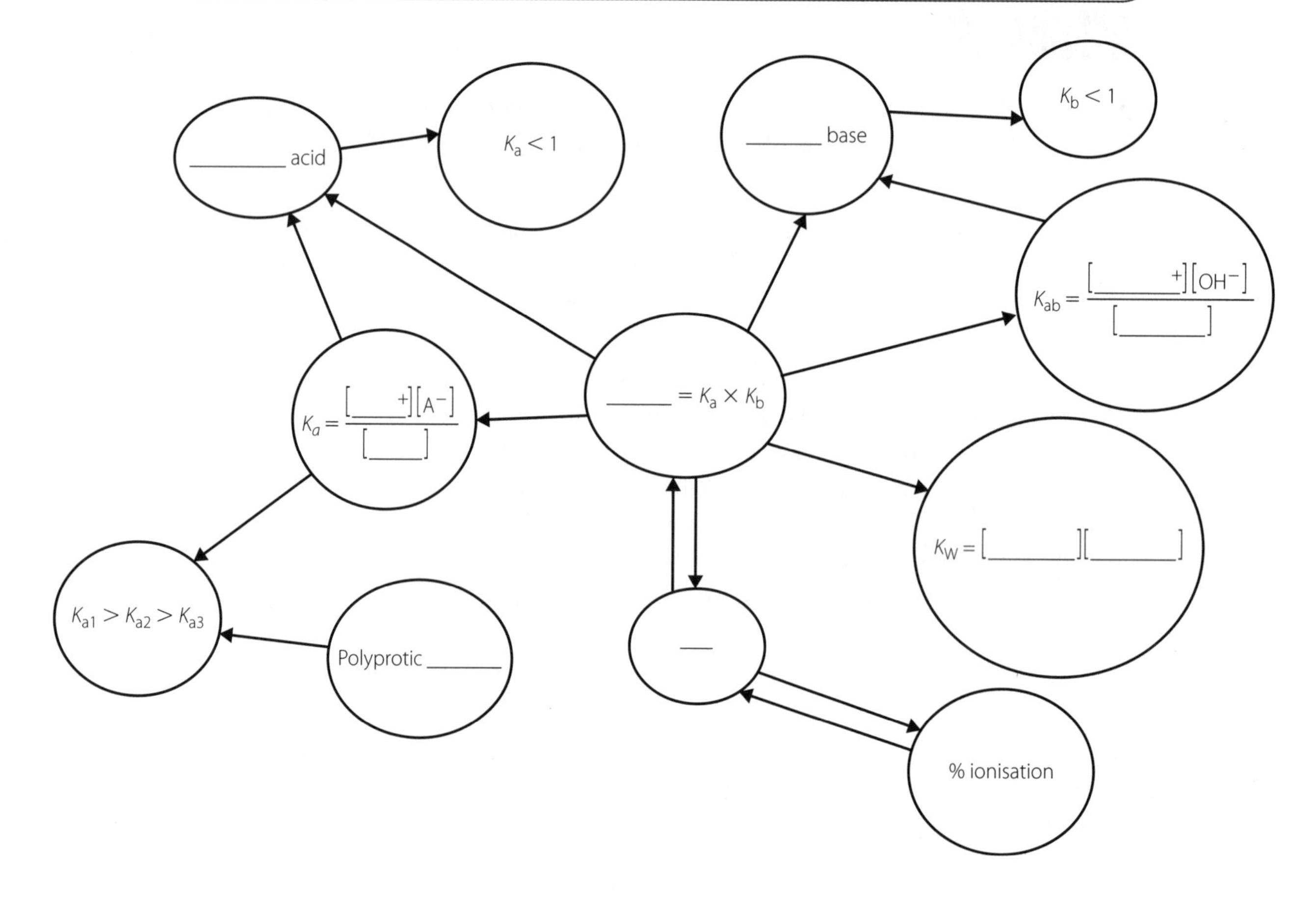

7.2 Acid dissociation constants

Weak acids only partially dissociate, setting up an equilibrium between the weak acid and its conjugate base:

$$HA(aq) + H_2O(l) \rightleftharpoons A^-(aq) + H_3O^+(aq)$$

The equilibrium expression for this general reaction is given by:

$$K_a = \frac{[H_3O^+][A^-]}{[HA]}$$

K_a is called the acid dissociation constant.

9780170412476

Ethanoic acid is a weak acid whose dissociation is given by:

$$CH_3COOH(aq) + H_2O(l) \rightleftharpoons CH_3COO^-(aq) + H_3O^+(aq)$$

When ethanoic acid dissociates, the amount of CH_3COO^- and H_3O^+ produced will be very small and so the function:

$$\frac{[CH_3COO^-][H_3O^+]}{[CH_3COOH]}$$

will give a very small value for K_a.

Similarly, weak bases only partially dissociate, setting up an equilibrium between the weak base and its conjugate acid:

$$B(aq) + H_2O(l) \rightleftharpoons HB^+(aq) + OH^-(aq)$$

The equilibrium expression is given by:

$$K_b = \frac{[HB^+][OH^-]}{[B]}$$

K_b is called the base dissociation constant.

Ammonia, NH_3, is a weak base:

$$NH_3(aq) + H_2O(l) \rightleftharpoons NH_4^+(aq) + OH^-(aq)$$

When ammonia dissociates, the amount of NH_4^+ and OH^- produced is very small and so the function:

$$K_b = \frac{[NH_4^+][OH^-]}{[NH_3]}$$

will give a very small value for K_b.

1 Fill in the blanks in the sentence below.

When ________ acids dissociate, an ____________________ is set up between the weak ______ and its conjugate ________. In this case, the equilibrium is shifted to the ________ because the amount of ____________ ____________ is very small compared to the amount of ____________. This gives rise to a very ________ K_a value.

2 For the substances listed below, decide whether they are acidic or basic, and write their dissociation equations and equilibrium expressions.

Dissociation equation	Substance	Acid or base	Equilibrium expression
______(aq) + $H_2O(l) \rightleftharpoons$ ______(aq) + ______(aq)	CH_3COOH		$K_{\_} = \frac{[\ \ ][\ \ ]}{[\ \ ]}$
______(aq) + $H_2O(l) \rightleftharpoons$ ______(aq) + ______(aq)	$CH_2ClCOOH$		$K_{\_} = \frac{[\ \ ][\ \ ]}{[\ \ ]}$
______(aq) + $H_2O(l) \rightleftharpoons$ ______(aq) + ______(aq)	$CH_3CH_2NH_2$		$K_{\_} = \frac{[\ \ ][\ \ ]}{[\ \ ]}$
______(aq) + $H_2O(l) \rightleftharpoons$ ______(aq) + ______(aq)	HBr		$K_{\_} = \frac{[\ \ ][\ \ ]}{[\ \ ]}$
______(aq) + $H_2O(l) \rightleftharpoons$ ______(aq) + ______(aq)	HN_3		$K_{\_} = \frac{[\ \ ][\ \ ]}{[\ \ ]}$

7.3 Analysing experimental data

An experiment is carried out on a number of acids and bases, P–Z. The substances are: hydrochloric acid, sulfuric acid, ethanoic acid (CH_3COOH), oxalic acid ($H_2C_2O_4$), sodium hydroxide, barium hydroxide, 1,3-propanediamine ($H_2NCH_2CH_2CH_2NH_2$) and methylamine (CH_3NH_2).

First, a few drops of universal indicator are added to each substance. Then, using a conductivity probe, the electrical conductivity of each substance is measured. The results are shown in the incomplete table below.

TABLE 7.3.1 Universal colour indicator and conductivity results for various acids and bases

SUBSTANCE	COLOUR IN UNIVERSAL INDICATOR	CONDUCTIVITY (μS)	IDENTIFICATION
P	Yellow	58	
Q	Purple	728	
R	Blue	62	
S	Red	665	
T	Purple	413	
X	Orange	83	
Y	Red	340	
Z	Dark blue	94	

Red	Pink	Orange	Beige	Yellow	Lime green	Green	Dark green	Turquoise	Pale blue	Blue	Dark blue	Violet	Purple

pH 1 2 3 4 5 6 7 8 9 10 11 12 13 14

1 Using the data given in the table, complete the identification column for each of the substances. Use their molecular name in your list.

2 In each case, provide evidence from the table to support your decision.

____

____

____

____

7.4 Calculations involving K_a and K_b

If the pH of a weak acid or base is known, then it is possible to determine its acid or base dissociation constant.

WORKED EXAMPLE 1

Calculate the K_a for ethanoic acid if a 0.1 M solution has a pH of 2.88.

ANSWER

a Write the dissociation equation for ethanoic acid.

$$CH_3COOH(aq) + H_2O(l) \rightleftharpoons CH_3COO^-(aq) + H_3O^+(aq)$$

b Write the expression for K_a.

$$K_a = \frac{[CH_3COO^-][H_3O^+]}{[CH_3COOH]}$$

c Calculate $[H_3O^+]$ using $pH = -\log_{10}[H_3O^+]$.

$$2.88 = -\log_{10}[H_3O^+]$$

$$[H_3O^+] = 1.32 \times 10^{-3}\,M$$

9780170412476

d Substitute the value for $[H_3O^+]$ into the K_a expression.
(Note: remember that at equilibrium, $[CH_3COO^-] = [H_3O^+]$.)

$$K_a = \frac{[H_3O^+]^2}{0.1}$$

$$K_a = \frac{(1.32 \times 10^{-3})^2}{0.1}$$

$$= 1.74 \times 10^{-5}$$

e It is very important when completing these calculations that any assumptions made are stated. The assumptions made were:

- At equilibrium, [HA] is the same as the initial concentration.
- $[H_3O^+]$ produced by the self-ionisation of water is negligible and has no effect on the calculations.

QUESTIONS

1 Calculate the K_a of an acid, HA, if a 0.125 M solution of the acid has a pH of 4.26.

____

____

____

____

____

2 Determine its K_b value of a base, B, if a 4.5×10^{-3} M solution of the base has a pH of 9.84.

____

____

____

____

____

____

CALCULATING pH FROM A GIVEN K_a VALUE

If the concentration of an acid or base is known, along with its K_a or K_b value, it is possible to calculate the pH.

WORKED EXAMPLE 2

Calculate the pH of a 0.096 M solution of nitrous acid, HNO_2 ($K_a = 4.5 \times 10^{-4}$).

ANSWER

a Write the dissociation equation for nitrous acid.

$$HNO_2(aq) + H_2O(l) \rightleftharpoons H_3O^+(aq) + NO_2^-(aq)$$

b Write the K_a expression.

$$K_a = \frac{[NO_2^-][H_3O^+]}{[HNO_2]}$$

c Substitute numbers in the equation from the question.

$$4.5 \times 10^{-4} = \frac{[NO_2^-][H_3O^+]}{0.096}$$

$$4.32 \times 10^{-5} = [H_3O^+]^2$$

$$6.57 \times 10^{-3}\,M = [H_3O^+]$$

d Substitute the value for $[H_3O^+]$ into the equation for pH.

$$pH = -\log_{10} 6.57 \times 10^{-3}$$

$$= 2.18$$

QUESTIONS

3 Calculate the pH of a 0.18 M solution of propanoic acid, CH_3CH_2COOH ($K_a = 1.34 \times 10^{-5}$).

4 Calculate the pH of a 0.05 M solution of ethylamine, $CH_3CH_2NH_2$ ($K_b = 5.95 \times 10^{-4}$).

CALCULATING PERCENTAGE IONISATION FROM A pH VALUE

Knowing the pH of an acid allows the $[H_3O^+]$ to be calculated. If the acid is monoprotic, then at equilibrium:

$$[H_3O^+] = [\text{conjugate base}]$$

This enables the percentage ionisation to be calculated according to the equation:

$$\text{Percentage ionisation} = \frac{[A^-]}{[HA]} \times 100\%$$

where $[A^-]$ is the concentration of the conjugate base at equilibrium and [HA] is the initial concentration of the acid.

WORKED EXAMPLE 3

A 1.78 M solution of hypochlorous acid, HOCl, has a pH of 3.65. Calculate the percentage ionisation of the acid.

ANSWER

a Write the dissociation equation for hypochlorous acid.

$$HOCl(aq) + H_2O(aq) \rightleftharpoons H_3O^+(aq) + OCl^-(aq)$$

9780170412476

b Using the pH value, calculate the $[H_3O^+]$.

$$3.65 = -\log_{10}[H_3O^+]$$
$$2.24 \times 10^{-4}\,M = [H_3O^+] = [OCl^-]$$

c Substitute $[H_3O^+]$ into the percentage ionisation equation.

$$\text{Percentage ionisation} = \frac{2.24 \times 10^{-4}}{1.78} \times 100\%$$
$$= 0.13\%$$

QUESTION

5 Calculate the percentage ionisation of a 0.89 M solution of an acid, HA, which has a pH of 2.79.

____

____

____

____

____

THE RELATIONSHIP BETWEEN K_a AND K_b

When a weak acid dissociates it produces a conjugate base that is also weak. It can be useful to be able to determine the strength of the conjugate base.

This is achieved using the equation:

$$K_w = K_a \times K_b$$

where K_w is the dissociation constant for the self-ionisation of water $= 1.0 \times 10^{-14}$.

WORKED EXAMPLE 4

Determine the strength of the ethanoate ion, CH_3COO^-, given that the K_a of ethanoic acid is 1.76×10^{-5}.

ANSWER

a Write the dissociation equation.

$$CH_3COOH(aq) + H_2O(l) \rightleftharpoons CH_3COO^-(aq) + H_3O^+(aq)$$

The conjugate base is the ethanoate ion, CH_3COO^-.

b Substitute the K_a value into the equation.

$$1.0 \times 10^{-14} = 1.76 \times 10^{-5} \times K_b$$
$$\frac{1.0 \times 10^{-14}}{1.76 \times 10^{-5}} = 5.68 \times 10^{-10}$$

QUESTION

6 Write the formula of the conjugate acid of pyridine, C_5H_5N, and determine its strength. (The K_b value for pyridine $= 1.5 \times 10^{-9}$.)

____

____

____

____

EVALUATION

1 The acid dissociation equilibrium constant:

A is represented by K_a.

B indicates the extent to which the acids accept a proton.

C is higher for weak acids than strong acids.

D is always 1 for strong acids.

2 The K_a of ethanoic acid is 1.8×10^{-5} whereas the K_a of hypochlorous acid is 3.5×10^{-8}. This indicates that:

A a greater percentage of hypochlorous acid will ionise than ethanoic acid.

B a $1\,mol\,L^{-1}$ solution of hypochlorous acid will produce more hydrogen ions than a $1\,mol\,L^{-1}$ solution of ethanoic acid.

C ethanoic acid is a stronger acid than hypochlorous acid.

D the pH of hypochlorous acid will be higher than that of ethanoic acid.

3 A student was comparing the electrical conduction of a variety of acids and bases. For the results to be valid the student would need to:

A use a new data probe for each solution.

B use the same concentration for each solution.

C test all the acids first, then the bases.

D use only weak acids or bases.

4 Which equation best describes how to calculate the percentage ionisation of an acid?

A $\%\text{ ionisation} = \dfrac{[H_3O^+]}{[A^-][H_3O^+]} \times 100$

B $\%\text{ ionisation} = \dfrac{[A^-]}{[HA]} \times 100$

C $\%\text{ ionisation} = \dfrac{[H_3O^+]}{[HA]} \times 100$

D $\%\text{ ionisation} = \dfrac{[A^-]}{[H_3O^+]} \times 100$

9780170412476

5 When calculating the acid dissociation constant from a pH value what assumptions need to be made? Explain why these assumptions are important.

6 When using conductivity values to rank acids and bases according to their strength, what other data needs to be gathered?

7 A 0.85 M solution of benzoic acid, C_6H_5COOH, has a pH of 2.13. Determine the strength of its conjugate base.

8 Acid–base indicators

Summary

- An indicator is a molecular substance that is used in titrations because it changes colour depending on the pH of the solution.
- An indicator is a weak acid and its conjugate base pair. The weak acid has a different colour from its conjugate base.
- The dissociation of an indicator can be represented by the general equation:

$$HIn + H_2O \rightleftharpoons In^- + H_3O^+$$

  where HIn and In– are the acid and base forms of the indicator respectively.
- An equilibrium expression can be written for the dissociation of an indicator:

$$K_{ind} = \frac{[In^-][H_3O^+]}{[HIn]}$$

- Indicators are chosen depending on their ability to change colour at or near the equivalence point of a titration.
- The equivalence point of a titration is the point when the reactants are present in the ratio shown by the mole ratio in the balanced chemical equation for the reaction.
- The end point of a titration is the physical sign that indicates that the equivalence point has been reached.
- At the equivalence point in a titration:

$$pK_{ind} = pH$$

8.1 Acid–base indicators

An acid–base indicator is a mixture of a weak acid and its conjugate base. The dissociation equation can be represented by the general equation:

$$HIn + H_2O \rightleftharpoons In^- + H_3O^+$$

The operation of an indicator is best understood by adopting Le Chatelier's principle:

- If a system at equilibrium is subject to a change in conditions, then the system will behave in such a way so as to partially counteract the change.

Consider methyl orange:

CH3, H3C, N+, H, N, N, O, OH — HIn, red $\underset{+H^+}{\overset{-H^+}{\rightleftharpoons}}$ CH3, H3C, N, N, N, O, O− — In−, yellow

If some methyl orange is placed into an acidic solution its equilibrium has been disturbed: it is surrounded by H_3O^+ ions. According to Le Chatelier, the system will move in the direction that will minimise the change; that is, it will shift in the direction that will decrease the concentration of H_3O^+ ions and it will shift left. In an acidic solution, methyl orange is red.

1 Explain, with reference to Le Chatelier's principle, why methyl orange is yellow in a basic solution.

____

____

____

____

____

____

____

____

8.2 The relationship between pH and pK_a

As with H_3O^+ concentration, the values for K_a values can be inconveniently small.

With H_3O^+ concentration is dealt with by expressing it as a pH calculated using the equation:

$$pH = -\log_{10} [H_3O^+]$$

Inconveniently small K_a (and K_b) values can be dealt with in the same way:

$$pK_a = -\log_{10} K_a$$

1 Fill in the blanks to complete the table.

ACID/BASE	K_a (or K_b)	pK_a (or pK_b)
Sulfurous acid	1.2×10^{-2}	
Propanoic acid		4.87
Ammonia		4.75
Ethylamine	5.37×10^{-4}	

2 If the dissociation of an indicator is given by:

$$HIn + H_2O \rightleftharpoons In^- + H_3O^+$$

fill in the blanks in the equilibrium expression:

$$K_{\_\_} = \frac{[\_\_\_][\_\_\_]}{[\_\_\_]}$$

3 At the equivalence point in a titration, the indicator is halfway through its colour change. The equilibrium above lies precisely in the middle. At this point:

$$[HIn] = [In^-]$$

Use the fact that $[HIn] = [In^-]$ at the equivalence point to fill in the blank in the equation. Explain your reasoning.

$$pK_{ind} = \_\_\_\_\_\_$$

____

____

____

4 Given that the useful pH range of an indicator is about ±1 of its pK_a value, complete the table below.

INDICATOR	K_a	pH RANGE
Methyl orange	2×10^{-4}	
Phenol red	1×10^{-8}	
Bromothymol blue	1×10^{-7}	
Phenolphthalein	5×10^{-10}	
Thymol blue	2×10^{-2}	
Litmus	1×10^{-7}	

8.3 Indicator colour change

The correct choice of indicator for an acid–base titration is extremely important, the colour change must occur very close to the equivalence point of the titration.

1 Fill in the blanks to complete the following sentences.
In the titration between hydrochloric acid and sodium hydroxide:

$$HCl(aq) + NaOH(aq) \rightleftharpoons NaCl(aq) + H_2O(l)$$

the ________________ point is when the exact amount of NaOH has been added to neutralise the HCl. The only substances present in the flask now are ____________ and ____________. Technically, the conjugate _________ of the HCl is ____________ but this has no discernible ________ properties because it is the conjugate _____________ of a ____________ acid.

Similarly, the conjugate acid of the NaOH is ___________ but this has no discernible _____________ properties because it is the conjugate ____________ of a ____________ base.

Therefore, the pH of the resultant solution is ___. A suitable indicator for this titration would be __________________________ because ____.

(Hint: refer to the table that accompanies question 4.)

WORKED EXAMPLE

Consider the titration between ammonia solution, $NH_3(aq)$, and HCl. Determine a suitable indicator for this titration.

ANSWER

a Write a balanced equation for the reaction.

$$HCl(aq) + NH_3(aq) \rightleftharpoons NH_4Cl(aq)$$

b Determine the species present at the equivalence point.
HCl(aq) has been used up.
No NH_3 present, only just enough was added to neutralise the HCl.
Only $NH_4Cl(aq)$ is present and since this is soluble, only the $NH_4^+(aq)$ ions and $Cl^-(aq)$ ions are present.

c Determine the acidic or basic nature of the species present.
Cl^- is the conjugate base of HCl but it has no basic properties.
NH_4^+ is the conjugate acid of a weak base and so will be a weak acid. As such, it should have a pH of about 5.

d Use an indicator chart (such as the one in your student book) to decide on the indicator.
A suitable indicator for this titration would be litmus.

QUESTION

2 Choose a suitable indicator for the titration between propanoic acid, CH_3CH_2COOH, and NaOH.

1 Indicators are not:

A varied in their K_a value.

B always red in acids.

C a solution in which at least one species has an intense colour.

D a solution of a conjugate acid–base pair.

2 A pH reading of _____ indicates a weak base.

A 4

B 7

C 12

D 9

3 An indicator changes colour in the pH range 3.8–5.4. What is the K_a of the indicator?

A 1.99×10^{-5}

B 5.01×10^{-10}

C 3.98×10^{-4}

D 5.01×10^{-8}

4 Which one of the following occurs in an indicator at the equivalence point of a titration?

A $HIn = [H_3O^+]$

B $[H_3O^+] = In^-$

C $HIn > In^-$

D $HIn = In^-$

5 Explain why the equivalence point is not at pH 7 when titrating ethanoic acid, CH_3COOH, with NaOH.

6 Alizarin yellow has a useful pH range between 10.2 (yellow) and 12.0 (red). Suggest an acid and a base titration for which this indicator would be useful.

7 With reference to Le Chatelier's principle, explain why the indicator bromothymol blue changes colour from blue in basic solution to yellow when acid is added.

9780170412476

8 The indicator phenol red changes colour from yellow in an acidic solution to red in a basic solution. Phenol red has an acid dissociation constant of 1×10^{-8}.

Determine the magnitude of the ratio $\frac{[In^-]}{[HIn]}$ for methyl red and the dominant species present in solutions of pH:

a 5

b 10.5

9 Volumetric analysis

Summary

- The equivalence point of a titration is the point when the reactants are present in the ratio shown by the mole ratio in the balanced chemical equation for the reaction.
- The end point of a titration is the physical sign that indicates that the equivalence point has been reached.
- There are four main types of acid–base titration: strong acid/strong base, strong acid/weak base, weak acid/strong base and weak acid/weak base. Weak acid/weak base titrations are not covered in the syllabus.
- A primary standard solution is one that can be made up to have an accurately known concentration that remains stable under common laboratory conditions for extended periods of time.
- In titrations the solution in the burette is called the *titrant* and the solution of unknown concentration is called the *analyte*.
- Back titration is a method used to determine the mass or concentration of a substance that is not able to be determined directly.
- When the use of indicators is not appropriate it is possible to monitor the progress of an acid–base titration using a pH probe. This produces a titration curve.
- The different types of titration produce distinctive titration curves.

9.1 Types of acid–base titration

STRONG ACID/STRONG BASE

For example, hydrochloric acid/sodium hydroxide:

$$HCl(aq) + NaOH(aq) \rightarrow NaCl(aq) + H_2O(l)$$

1 Complete the sentences by filling in the blanks.
At the ________________ point of a strong acid (HCl)/strong base (NaOH) titration, just enough NaOH has been added to neutralise the ____________. The solution remaining contains _________ ions, __________ ions and ___________. The pH of this solution is _______ because ____ ________________________. An indicator with a useful pH range between ________ and _________ would be suitable for this type of titration.

STRONG ACID/WEAK BASE

For example, hydrochloric acid/NH_3:

$$HCl(aq) + NH_3(aq) \rightarrow NH_4Cl(aq)$$

2 Complete the sentences by filling in the blanks.
At the ________________ point of a strong acid (HCl)/weak base (NH_3) titration, just enough NH_3 has been added to neutralise the ____________. The solution remaining contains _________ ions and __________ ions. The pH of this solutions is _______ because ____ ____. An indicator with a useful pH range between ________ and _________ would be suitable for this type of titration.

WEAK ACID/STRONG BASE

For example, ethanoic acid, CH_3COOH/NaOH:

$$CH_3COOH(aq) + NaOH(aq) \rightarrow CH_3COONa(aq) + H_2O(l)$$

3 Complete the sentences by filling in the blanks.
At the ________________ point of a weak acid (CH_3COOH)/ strong base (NaOH) titration, just enough NaOH has been added to neutralise the ____________. The solution remaining contains _________ ions, ___________ ions and ___________. The pH of this solution is _______ because _______________________________________ ____. An indicator with a useful pH range between ________ and _________ would be suitable for this type of titration.

WEAK ACID/WEAK BASE

For example, ethanoic acid, CH_3COOH/NH_3:

$$CH_3COOH(aq) + NH_3(aq) \rightleftharpoons CH_3COO^-(aq) + NH_4^+(aq)$$

This is a complex system with no clear-cut equivalence point. An acid–base indicator would not be appropriate to monitor this type of reaction.

9.2 Acid–base titrations

THE PRIMARY STANDARD

In volumetric analysis, particularly acid–base titrations, the concentration of one of the reactants must be accurately known. This is not possible with common reactants such as NaOH and HCl whose concentration can change significantly over time.

In order to use these reactants in titrations, their concentrations must be determined by titrating them initially with a primary standard. Primary standards are substances whose purity and stability enable them to be made into solutions whose concentration can be accurately known.

Common primary standards used in acid–base titrations include:

- potassium hydrogen phthalate (KHP), $KH(C_8H_4O_4)$. This is a monoprotic weak acid that is available in solid powder form.
- sodium carbonate, anhydrous (Na_2CO_3). Available in solid powder form.

WORKED EXAMPLE 1

A potassium hydrogen phthalate (KHP) solution was prepared by dissolving 4.216 g of KHP in 50 mL of distilled water. This was transferred, with washing, to a 250 mL volumetric flask and made up to the mark with distilled water. A 20 mL sample of this solution was transferred to a conical flask. A few drops of phenolphthalein indicator were added to the KHP solution and this was titrated with a roughly 0.1 M solution of NaOH. The pale pink end point occurred when 16.80 mL of NaOH had been added.

Determine the concentration of the NaOH to three decimal places.

ANSWER

a Calculate the number of moles of KHP used.

$$n = \frac{m}{M} = \frac{4.216}{204.22} = 0.0206\,\text{mol}$$

0.0206 mol of KHP was dissolved in water and then transferred to the 250 mL volumetric flask.
A 20 mL sample of this solution was transferred to the conical flask.

b Calculate the number of moles of KHP in the conical flask.
20 mL is 12.5 times smaller than 250 mL. Therefore, the number of moles of KHP in 20 mL is 12.5 times smaller than in 250 mL.

$$\frac{20.00}{250.00} = \frac{n\text{ in }20.00}{\mathbf{0.0206}} \quad n\text{ in }20\,\text{mL}$$

$$= 1.648 \times 10^{-3}\,\text{mol}$$

c Write the equation for the reaction to determine the mole ratio between KHP and NaOH.

$$KH(C_8H_4O_4)(aq) + NaOH(aq) \rightarrow Na^+(aq) + K(C_8H_4O_4)^-(aq) + H_2O(aq)$$

1 mol KHP = 1 mol NaOH

d Use the mole ratio to determine the number of moles of NaOH added from the burette.

1.648×10^{-3} KHP in the conical flask

$= 1.648 \times 10^{-3}$ mol NaOH added from burette

e Determine the concentration of NaOH from the end point volume of 16.80 mL.

$$c = \frac{n}{V}$$

$$= \frac{1.648 \times 10^{-3}}{\frac{16.80}{1000}}$$

$$= 0.098\,\text{M}$$

9780170412476

QUESTION

1 A standard solution of sodium carbonate, Na_2CO_3, was prepared by dissolving 1.624 g of sodium carbonate in 50 mL of distilled water. This was transferred, with washing, to a 250 mL volumetric flask and made up to the mark with distilled water. A pipette was used to transfer 20 mL of this solution to a conical flask. A few drops of methyl orange were added and the solution titrated with HCl until the pale orange end point was reached. The volume of HCl added from the burette was 15.65 mL.

Determine the concentration of HCl.

It can be useful to visualise these questions by drawing a diagram and calculating each of the steps in the titration.

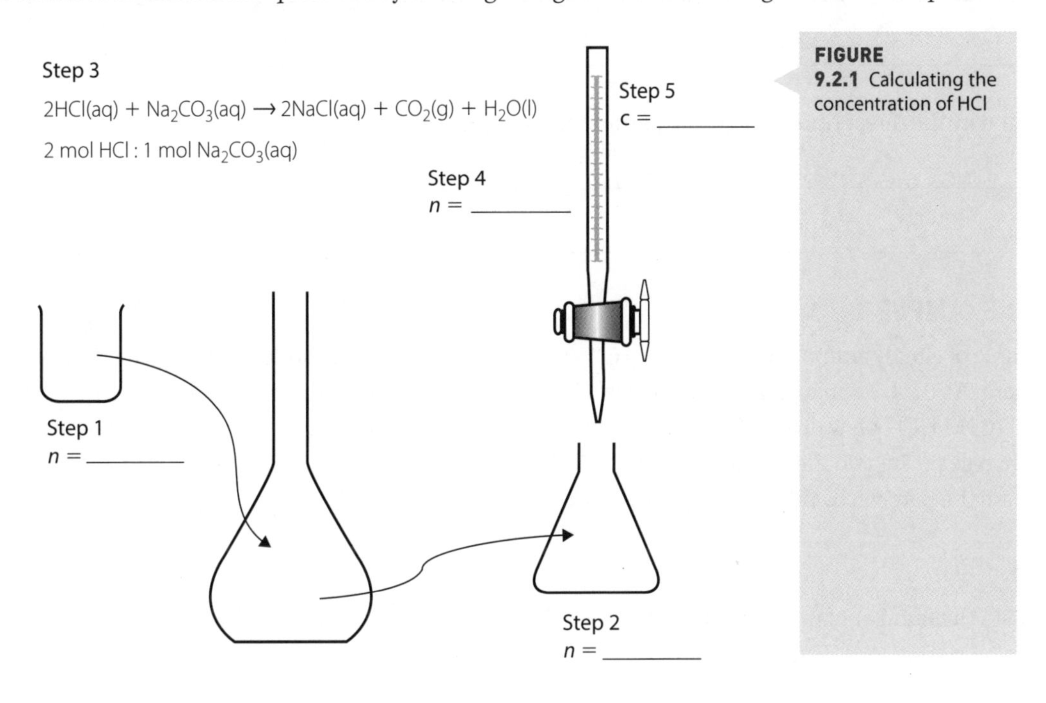

FIGURE 9.2.1 Calculating the concentration of HCl

____

____

____

____

____

____

____

____

____

____

BACK TITRATIONS

2 When determining the amount of ammonia in a commercially available brand of cloudy ammonia it is necessary to perform a back titration. Explain why it is sometimes necessary to perform a back titration.

____

____

____

____

3 Explain why the direct titration of cloudy ammonia is not possible.

____

____

WORKED EXAMPLE 2

A 10 g sample of cloudy ammonia was added to a 250 mL volumetric flask. Distilled water was added up to the 250 mL mark. A 20 mL sample of this solution was transferred by a pipette to a conical flask where it was mixed with 20 mL of 0.103 M HCl. A few drops of phenolphthalein indicator were added and the mixture was titrated with 0.052 M NaOH. The pale pink end point occurred when 6.50 mL of NaOH had been added.

Calculate the percent, by mass, of NH_3 in the sample of cloudy ammonia.

ANSWER

a Calculate the number of moles of HCl added initially in the 20 mL of 0.103 M.

$$n_{HCl} = c \times V$$

$$= \_\_\_\_ \times \frac{\_\_\_\_}{1000}$$

$$= \_\_\_\_\_\_ \text{ mol (initial)}$$

b Calculate the number of moles of NaOH added from the burette.

$$n_{NaOH} = c \times V$$

$$= \_\_\_\_ \times \frac{\_\_\_\_}{1000}$$

$$= \_\_\_\_\_\_ \text{ mol}$$

c Write the equation for the reaction between HCl and NaOH to determine the number of moles of HCl in the conical flask.

$$HCl(aq) + NaOH(aq) \rightarrow NaCl(aq) + H_2O(l)$$

The mole ratio is 1 mol HCl = 1 mol NaOH.

$$n_{HCl} = \_\_\_\_\_\_ \text{ mol (final)}$$

Step **a** calculated the number of moles of HCl added *initially* to the ammonia solution. The number of moles of HCl is deliberately chosen to be in *excess* so that some HCl reacts with all of the NH_3 according to the equation:

$$HCl(aq) + NH_3(aq) \rightarrow NH_4Cl(aq)$$

and some HCl is leftover to be titrated with the NaOH.

Step **b** calculated the number of moles of NaOH that was equal to the number of moles of HCl in the conical flask.

Step **c** calculated the number of moles of HCl in the conical flask. This is the number of moles of HCl left over from the initial amount after reaction with NH_3.

d Calculate the number of moles of HCl that reacted with the NH_3.

$$n_{\text{HCl reacted}} = n_{\text{HCl initial}} - n_{\text{HCl final}}$$
$$= \_\_\_\_\_\_\_\_\_\_ - \_\_\_\_\_\_\_\_\_\_$$
$$= \_\_\_\_\_\_\_\_\_\_ \text{ mol (reacted)}$$

e Write the equation for the reaction between HCl and NH_3 to determine the mole ratio and hence the number of moles of NH_3 in the sample.

$$HCl(aq) + NH_3(aq) \rightarrow NH_4Cl(aq)$$
$$1 \text{ mol HCl} = 1 \text{ mol } NH_3$$

If there were ____________ mol HCl (reacted) there must have been ____________ mol NH_3 (present).

f Calculate the number of moles of NH_3 in the cloudy ammonia.
n NH_3(present) in the conical flask is 12.5 times smaller than in the 250 mL volumetric flask.

$$\text{Therefore, } n(NH_3) \text{ in 250 mL} = \_\_\_\_\_\_\_\_\_\_ \text{ mol } NH_3 \text{ (present)} \times 12.5$$
$$= \_\_\_\_\_\_\_\_\_\_ \text{ mol } NH_3 \text{ in 250 mL}$$

This is the number of moles of NH_3 in the sample of cloudy ammonia.

g Calculate the mass of NH_3 in the cloudy ammonia.

$$m(NH_3) = \text{mol } NH_3 \text{ in 250 mL} \times 17$$
$$= \_\_\_\_\_\_\_\_\_\_ \text{ g } NH_3$$

h Calculate the percentage of NH_3 in the sample.

$$\% \, NH_3 = \frac{\_\_\_\_\_\_\_\_\_\_}{10.00} \times 100$$
$$= \_\_\_\_\_\_\_\_\_\_ \, \%$$

QUESTION

4 An investigation to determine the amount of calcium ions, Ca^{2+}, in a sample of limestone was carried out. A 1.336 g sample of limestone was crushed to a powder and dissolved in 100 mL of 0.173 M HCl in a conical flask. The solution was titrated with 0.114 M NaOH using a few drops of phenolphthalein as the indicator. The end point occurred at 21.60 mL. Determine the percentage of Ca^{2+} ions in the limestone.

9.3 Graphs

The use of an indicator to mark the equivalence point in an acid–base titration is not always possible; for example, the acid, the base or both could be strongly coloured. In these situations a pH probe can be used.

The equivalence point can be determined by monitoring the pH of the titration as it progresses. The data gathered takes the form of a graph of pH against volume of acid or base added.

Given that there are four different types of acid–base titration, each type produces its own unique graph called a titration curve.

The different types of titrations curves are shown below.

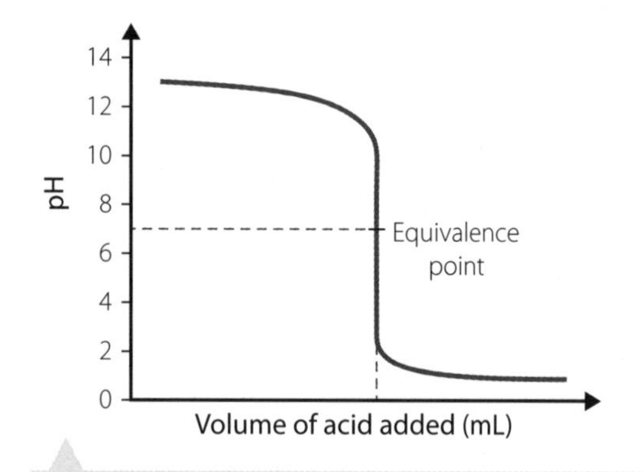

FIGURE 9.3.1 Titration curve for a strong acid/strong base

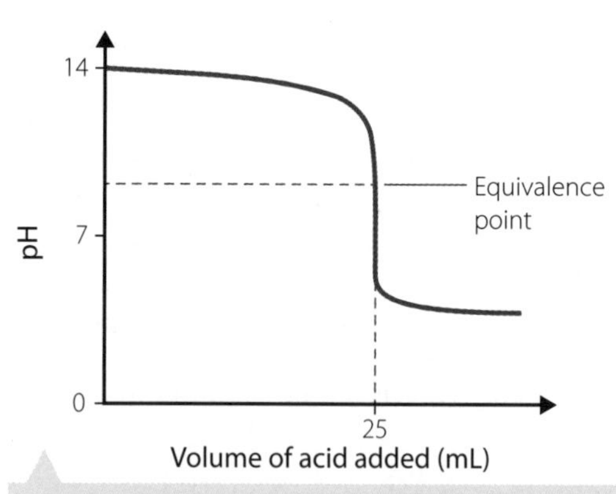

FIGURE 9.3.2 Titration curve for a weak acid/strong base

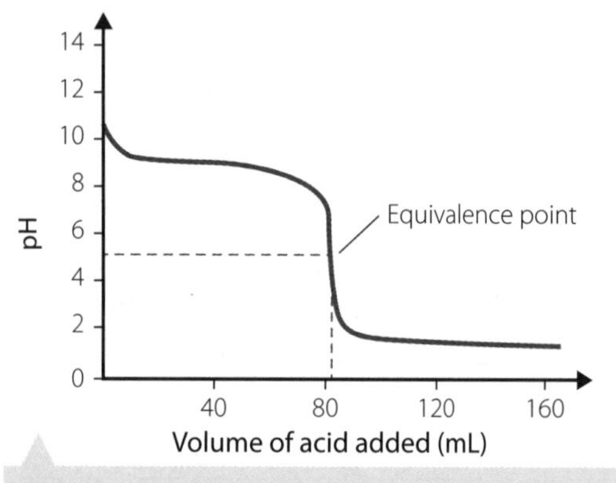

FIGURE 9.3.3 Titration curve for a strong acid/weak base

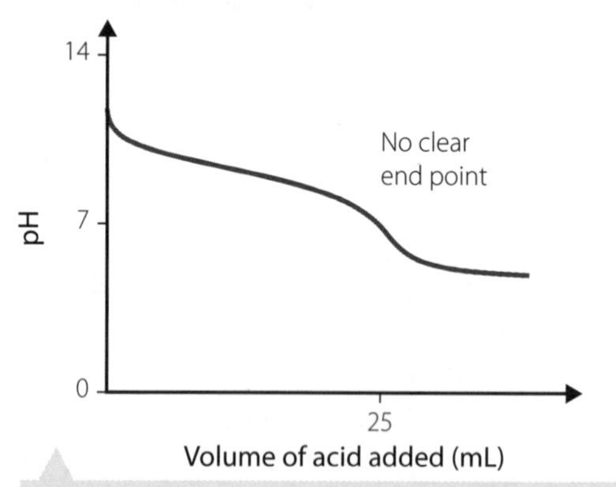

FIGURE 9.3.4 Titration curve for a weak acid/weak base

1 Complete the following titration curve.

FIGURE 9.3.5 Titration curve for __________ base/__________ acid

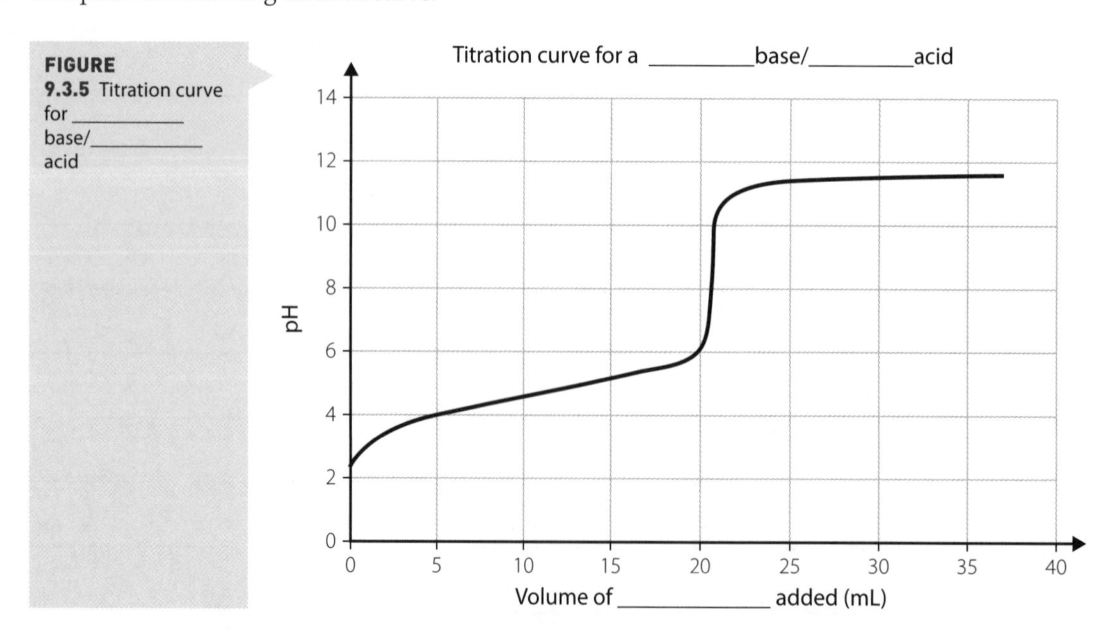

9780170412476

2 Add the following terms to the graph in question 1.

Buffer region	Equivalence point
Initial pH	$pH = pK_a$

3 Suggest a suitable acid and base for this titration and write an explanation for the pH of the equivalence point.

1 When a solution of sodium hydroxide reacts with ethanoic acid, assuming there are no excess reactants, the resultant solution will be:

A weakly acidic.

B highly alkaline.

C neutral.

D weakly alkaline.

2 When measuring the volume of a solution the most accurate method would be to use a:

A measuring cylinder.

B pipette.

C beaker.

D conical flask.

3 On a titration curve for a weak acid/strong base, the equivalence point is about pH 9. This is because:

A the only remaining species present is a conjugate acid.

B $[H_3O^+] > [OH^-]$.

C the only remaining species present is a conjugate base.

D the reaction between the acid and base is not complete.

4 Which one of the following is *not* a desirable property of a primary standard?

A It has a small molar mass so that small amounts can be easily measured.

B It is available in a highly pure form.

C It must be stable in the presence of air.

D It should be cheap.

9780170412476

5 Explain why a back titration would be necessary to determine the amount of $Mg(OH)_2$ in an antacid tablet.

6 Research the term 'concordant titre'. Define and explain why it is important.

7 A 0.622 g sample of chalk was dissolved in 30 mL of 0.317 M HCl. The mixture was titrated with 0.04 M NaOH. The end point occurred when the phenolphthalein indicator turned pale pink after 19.80 mL of NaOH had been added. Determine the percentage of calcium carbonate, by mass, in the sample of chalk.

10 Redox reactions

Summary

- Many types of reactions can be modelled as redox reactions involving the oxidation of one substance and the reduction of another substance. These reactions include:
  - metal displacement reactions
  - combustion reactions
  - corrosion reactions
  - electrochemical processes.
- Oxidation can be defined as the *loss* of electrons.
- Reduction can be defined as the *gain* of electrons.
- Metals can be ordered according to their strength as reducing agents. This ordering of metals is referred to as the activity series of metals.
- The reducing strength (reactivity) of a metal can be predicted by the position of the metal in the periodic table.
- Oxidation and reduction processes are complimentary and can be represented using half-equations.
- Complex redox reactions in acidic solutions can be balanced by balancing each oxidation and reduction half-equation separately, and then balancing the electrons lost and gained.
- The oxidation state of an element in a compound or ion can be determined by following a specific set of rules.
- The oxidation state of a substance represents its 'charge' in a compound.

9780170412476

10.1 Concept map

The following shows a concept map for redox reactions. Some of the concepts have been provided, the rest are given in the concept list. Use the list to complete the concept map.

Concept list

$2Fe(OH)^2(s)$	$CO_2(g) + H_2O(l)$	combustion	corrosion
decrease	electrolytic	gain	galvanic
increase	loss	$M^{2+}(aq) + X(s)$	metal
oxidation	reduction		

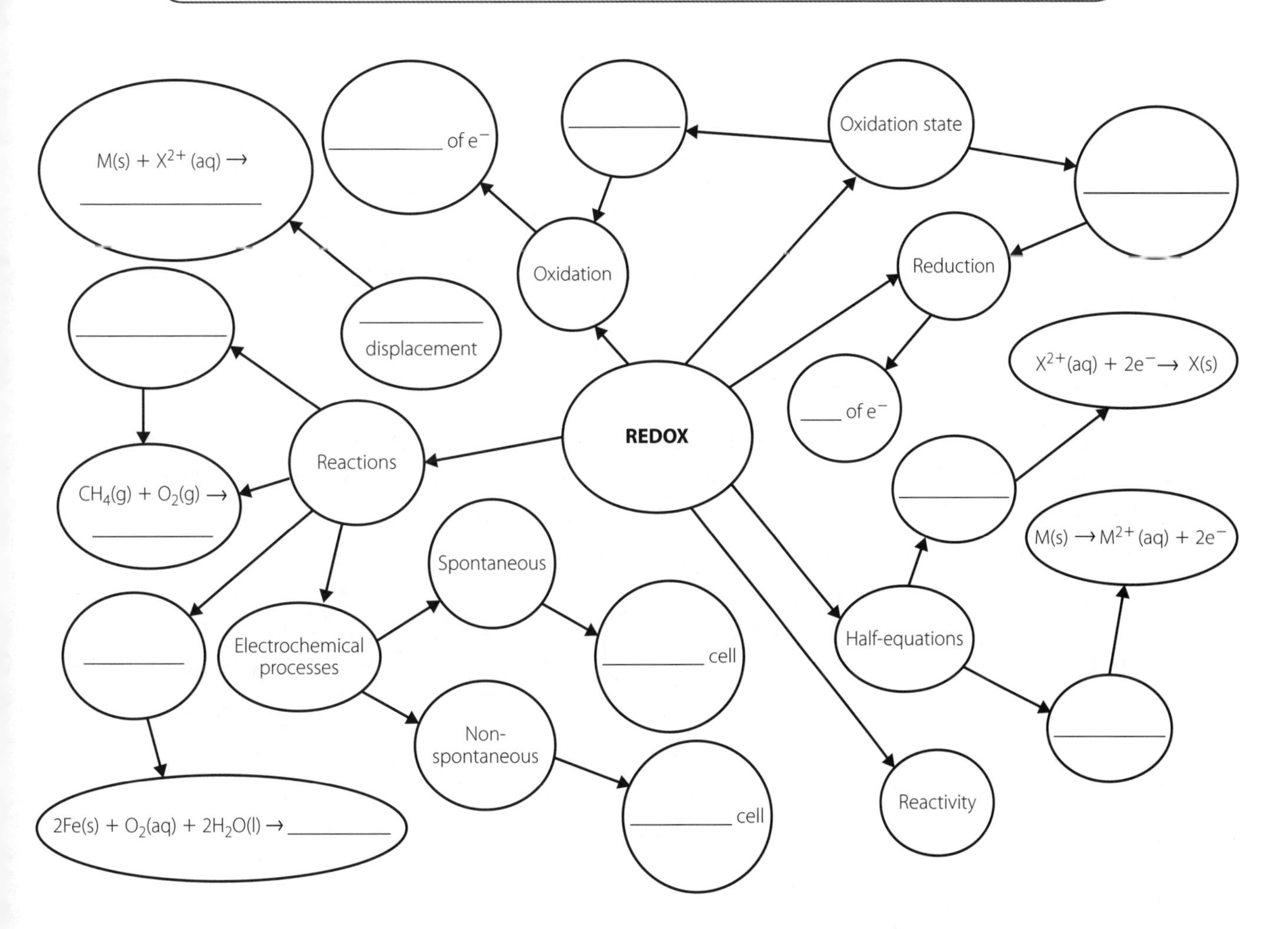

10.2 Range of redox reactions

Redox reactions are oxidation and reduction reactions, and involve the transfer of electrons. A number of reaction types can be classified as redox, including: metal displacement, combustion and corrosion reactions.

METAL DISPLACEMENT REACTIONS

The general form of a metal displacement reaction is given by:

$$AB(aq) + C(s) \rightarrow CB(aq) + A(s)$$

where

- AB = solution of a metal compound
- C = metal
- CB = solution of the displacing metal compound
- A = displaced metal

For example:

$$Zn(s) + CuSO_4(aq) \rightarrow ZnSO_4(aq) + Cu(s)$$

The oxidising and reducing species can be identified more easily if the net ionic equation is written. This is done by removing the spectator ion, SO_4^{2-}:

$$Zn(s) + Cu^{2+}(aq) \rightarrow Zn^{2+}(aq) + Cu(s)$$

By tracking the outcome of the reaction with respect to each reactant:

$$Zn(s) \rightarrow Zn^{2+}(aq) + 2e^-$$

The zinc has *lost* electrons. It has been *oxidised.*

$$Cu^{2+}(aq) + 2e^- \rightarrow Cu(s)$$

The copper has *gained* electrons. It has been *reduced.*

These are examples of oxidation and reduction half-equations. These are very important to the understanding of redox chemistry.

In the examples described above, Zn is more reactive than Cu because it is able to displace Cu^{2+} ions from $CuSO_4$. If the reverse was attempted:

$$Cu(s) + Zn^{2+}(aq) \rightarrow \text{no reaction}$$

In this way, an activity series of metals can be produced.

1 A student carried out a series of experiments on a number of unknown metals. On her laboratory bench she had four unknown metals labelled Q, T, L and D. She also had solutions of the nitrate of each metal, a spotting plate and dropping pipettes. Figure 10.2.1 shows the results of the experiments.

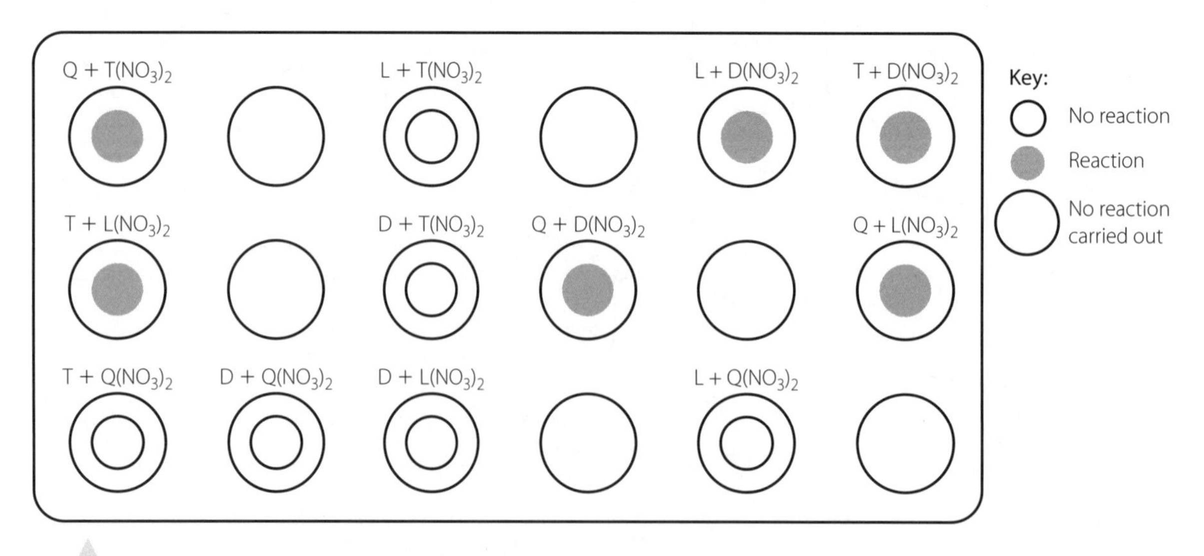

FIGURE 10.2.1 Experimental results

Using the information given, list the metals in order of activity from strongest oxidising agent to weakest oxidising agent. Show all working with the aid of net ionic, oxidation and reduction half-equations.

COMBUSTION REACTIONS

A combustion reaction is defined as an exothermic reaction involving a fuel and an oxidant, usually gaseous oxygen. For example, the reaction between magnesium and oxygen is a highly exothermic reaction that produces an intense white light:

$$2Mg(s) + O_2(g) \rightarrow 2MgO(s)$$

Writing the half-equations gives:

$$Mg(s) \rightarrow Mg^{2+}(s) + 2e^- \quad \textit{Oxidation}$$

$$O_2(g) + 4e^- \rightarrow 2O^{2-}(s) \quad \textit{Reduction}$$

2 Construct oxidation and reduction half-equations for the combustion of Na, Ca, and Al.

CORROSION REACTIONS

Corrosion reactions refer to the corrosion or rusting of iron.

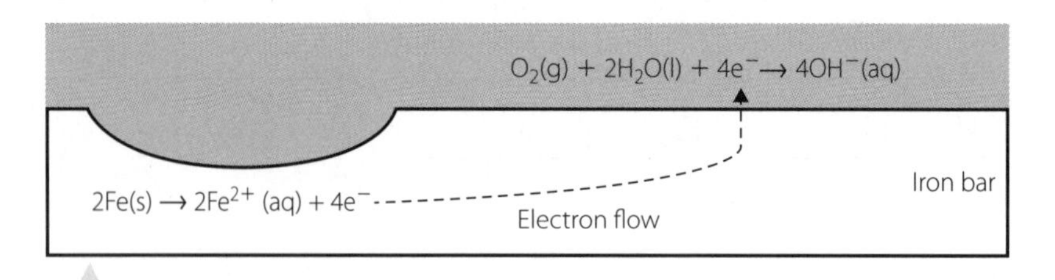

FIGURE 10.2.2 Oxidation and reduction processes involved in the corrosion of an iron bar

3 a State the conditions required for the corrosion of iron to occur.

b Suggest two ways that an iron bar could be protected from corrosion.

____

____

____

____

____

____

c Corrosion can be thought of as an electrochemical process. Figure 10.2.2 is a real-life representation of a galvanic cell. Add labels to the figure to show the anode, cathode and electrolyte.

PREDICTING LOSS AND GAIN OF ELECTRONS

Metallic elements tend to lose electrons in order to become chemically stable. As a result, they form positively charged ions or *cations*.

Non-metallic elements tend to gain electrons in order to become chemically stable. As result, they form negatively charged ions or *anions*.

When predicting the ease in which metallic elements lose their electrons, many factors need to be considered. For instance, the:

- number of protons in the nucleus: more protons = stronger attraction for valence electrons
- distance between valence electrons and nucleus: greater the distance = less strongly attracted
- number of shells between nucleus and valence electrons: more shells = less strongly attracted.

4 a Use the information in Figure 10.2.3 and Table 10.2.1 to complete the activity series in Figure 10.2.4.

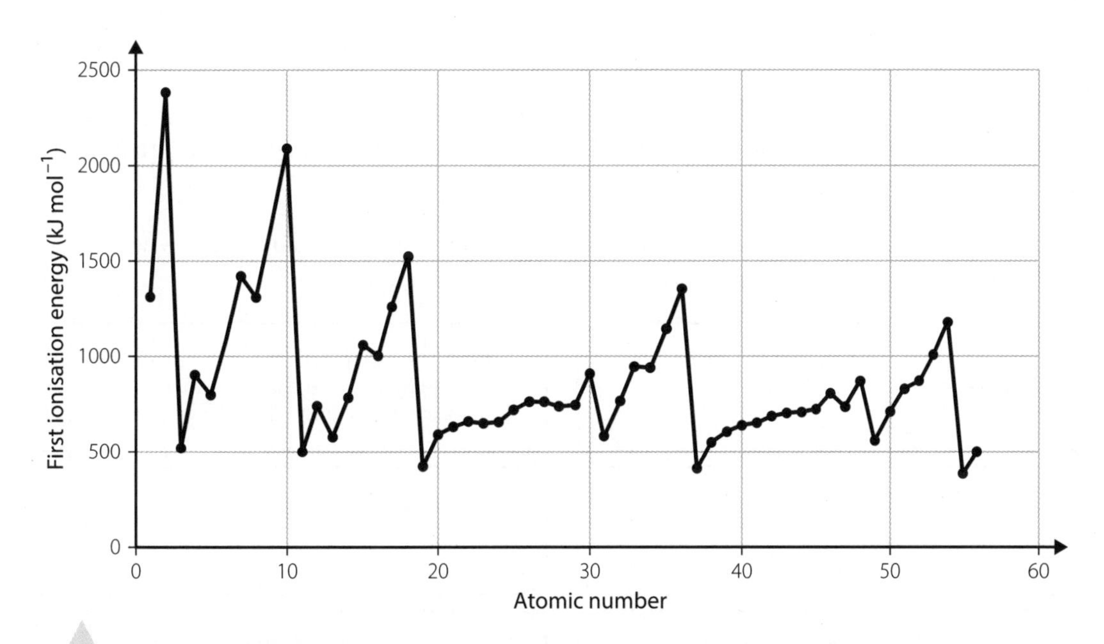

FIGURE 10.2.3 Graph of first ionisation energy against atomic number, for elements 1–56

TABLE 10.2.1 A selection of elements and their atomic numbers

LETTER	A	B	C	D	E	F	G	H	I	J	K	L	M	N	O
ATOMIC NUMBER	12	37	11	18	56	20	2	3	55	38	10	25	36	54	8

☐ < ☐ < ☐ < ☐ < ☐ < ☐ < ☐ < ☐ < ☐ < ☐

FIGURE 10.2.4 Activity series for a selection of elements

b Table 10.2.1 gives the atomic number of fifteen elements, yet there are only ten boxes available in which to place letters. Which elements were left out of the activity series?

____

c Explain why these elements were left out.

____

____

d From the information given, why would it be difficult to rank the activities of elements 22–29?

____

____

____

____

BALANCING REDOX EQUATIONS IN ACIDIC SOLUTIONS

Writing balanced equations for redox reactions in acidic solutions can be complicated but can be achieved by using a set of rules.

5 Write a balanced equation for the reaction between potassium permanganate and ethanol in acidic solution. Use the following rules to complete the flowchart and write the balanced equation.

Rule 1: Charge is balanced by adding the appropriate number of electrons to the more positive side of the equation.

Rule 2: If the number of electrons in a half-equation is multiplied by a number, then everything in the half-equation must be multiplied by the same number.

Rule 3: Write the partial or skeleton half-equation.

Rule 4: Balance for elements other than O and H.

Rule 5: Balance for O by adding the appropriate number of H_2O molecules to the opposite side.

Rule 6: Balance for H by adding the appropriate number of H^+ ions to the opposite side.

9780170412476

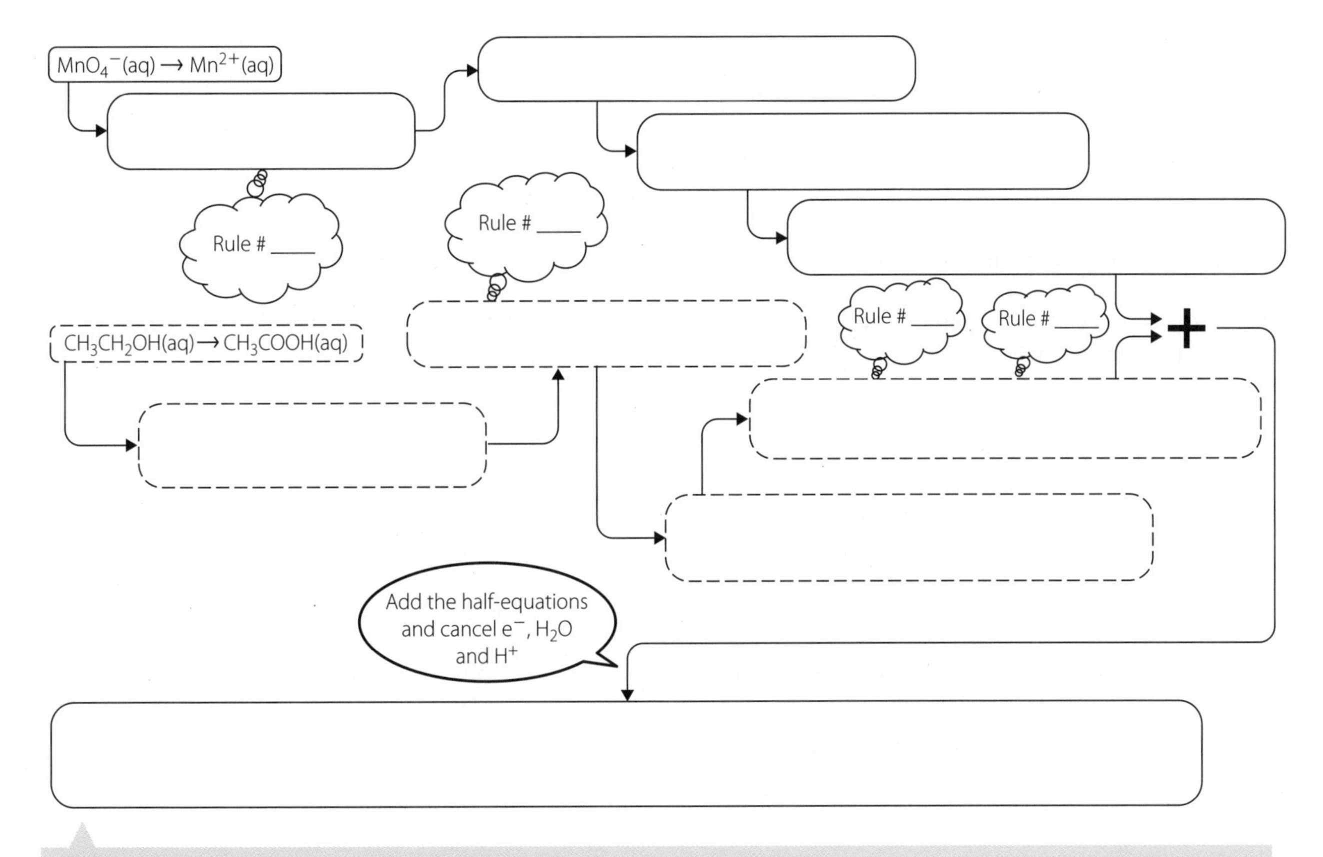

FIGURE 10.2.5 Constructing the balanced equation

DEDUCING OXIDATION STATES

Deducing the oxidation state of an element in a compound or a polyatomic ion is very important when attempting to show whether a given reaction is a redox reaction or not.

For example, is the reaction between acidified dichromate and oxalate a redox reaction?

$$14H^+(aq) + Cr_2O_7^{2-}(aq) + 3C_2O_4^{2-}(aq) \rightarrow 2Cr^{3+}(aq) + 6CO_2(aq) + 7H_2O(l)$$

The oxidation state changes of the reactants (if any) need to be tracked using the rules for assigning oxidation states given in Figure 10.2.6.

1 The oxidation state of any element is zero.

2 The oxidation state of a monatomic ion is the same as the charge on that ion.

3 The oxidation state of hydrogen in compounds is always +1, except when forming metal hydrides when it is −1.

4 The oxidation state of oxygen in any compound is always −2, except when forming peroxides where it is −1 and in F_2O where it is +2.

5 In a neutral compound, the sum of the oxidation states must always equal zero.

6 In ions, the sum of the oxidation states must always equal the charge on the ion.

FIGURE 10.2.6 Rules for assigning oxidation states.

Considering chromium, Cr:

$$Cr_2O_7^{2-} \longrightarrow Cr^{3+}$$

$7 \times -2 = -14$

$-14 + Cr_2 = -2$

$Cr_2 = +12$

Cr = +6

Cr = +3

Cr has changed from +6 to +3, which is a decrease in oxidation state. Cr has been *reduced.*

6 Complete the calculations for the carbon in oxalate. If the chromium has been reduced, then the carbon must have been oxidised. Verify that this is the case and decide if the reaction is redox.

7 Using the approach outlined above, decide whether the following reactions are redox reactions.

a $3Ca(HCO_3)_2(aq) + 2H_3BO_3(aq) \rightarrow Ca_3(BO_3)_2(aq) + 6CO_2(g) + 6H_2O(l)$

b $NH_4Cl(aq) + NaOH(aq) \rightarrow NH_3(aq) + NaCl(aq) + H_2O(l)$

c $2C_8H_{18}(l) + 25O_2(g) \rightarrow 16CO_2(g) + 18H_2O(l)$

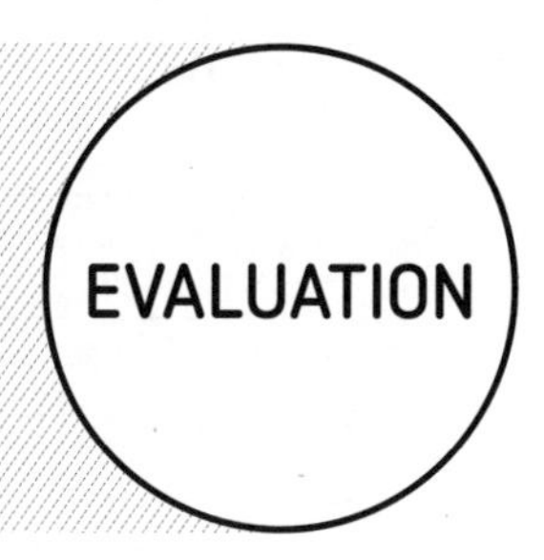

1 When a piece of iron corrodes:

A the iron is reduced and so is the anode in the electrochemical process.

B the iron is oxidised and so is the anode in the electrochemical process.

C the iron is oxidised and so is the cathode in the electrochemical process.

D the iron is reduced and so is the cathode in the electrochemical process.

2 When permanganate ions in acidified solution react, the manganese becomes:

A Mn(V) ions.

B Mn(III) ions.

C Mn(VII) ions.

D Mn(II) ions.

3 In the reaction between copper(II) oxide and carbon monoxide, the reducing agent is:

A Cu.

B CO.

C Cu^{2+}.

D CO_2.

4 The combustion of ethene, C_2H_4, in oxygen is a redox reaction because:

A the oxygen is oxidised and the carbon is reduced.

B the carbon is the reducing agent and oxygen is the oxidising agent.

C the oxidation state of carbon decreases and that of oxygen increases.

D the carbon gains electrons and the oxygen loses electrons.

5 Determine the oxidation state of the **bold** element in the following questions.

a $\mathbf{Cr}O_4^-$

b $\mathbf{C}_2O_4^{2-}$

c BO_3^{3-}

d S_8^-

6 List the three factors that need to be considered when determining the reactivity of a metal.

7 Refer to the activity series completed in section 10.2, question 4. In a galvanic cell constructed using elements A and L, state which metal is the oxidising species and which metal is the reducing species. Include the appropriate half-equations to support your answer.

9780170412476

8 When a potassium permanganate solution is used in volumetric analysis, it first has to be standardised in order to determine its exact concentration. This involves titrating it with a primary standard such as sodium oxalate. Given the skeleton half-equations, write a balanced equation for the reaction between acidified potassium permanganate and sodium oxalate.

$$MnO_4^-(aq) \rightarrow Mn^{2+}(aq)$$

$$C_2O_4^{2-}(aq) \rightarrow CO_2(g)$$

11 Electrochemical cells

Summary

- Electrochemical cells allow for the transformation of energy between chemical potential energy and electrical energy.
- Galvanic cells transform chemical potential energy into electrical energy.
- Electrolytic cells transform electrical energy into chemical potential energy.
- Spontaneous metal displacement reactions are examples of a redox reaction. The electrons produced by the oxidised metal can be harnessed to do work in a galvanic (or voltaic) cell.
- A non-spontaneous reaction can be forced to occur by the application of an electric current. This is the basis of an electrolytic cell.
- In both kinds of electrochemical cell, oxidation occurs at the anode.
- In galvanic cells, the two half-cells are connected by a salt bridge. This, together with the electrodes and the electrolytes, make up the internal circuit.
- The wires, and any meters and/or appliances, make up the external circuit.

9780170412476

11.1 Electrolytic cells

1 Use the word list to assign correct labels on the two types of electrochemical cells: a galvanic cell (Figure 11.1.1) and an electrolytic cell (Figure 11.1.2).

$2Br^-(l) \rightarrow Br_2(g) + 2e$	anode	anode	$Br^-(l)$
electron flow	cathode	cathode	electron flow
$K^+(aq)$	lead	$Li^+(l)$	$Li^+(l) + e^- \rightarrow Li(l)$
$Ni^{2+}(aq)$	nickel	$Ni(s) \rightarrow Ni^{2+}(aq) + 2e^-$	
NO_3^-	platinum electrodes	$Pb^{2+}(aq)$	
$Pb^{2+}(aq) + 2e^- \rightarrow Pb(s)$	salt bridge		

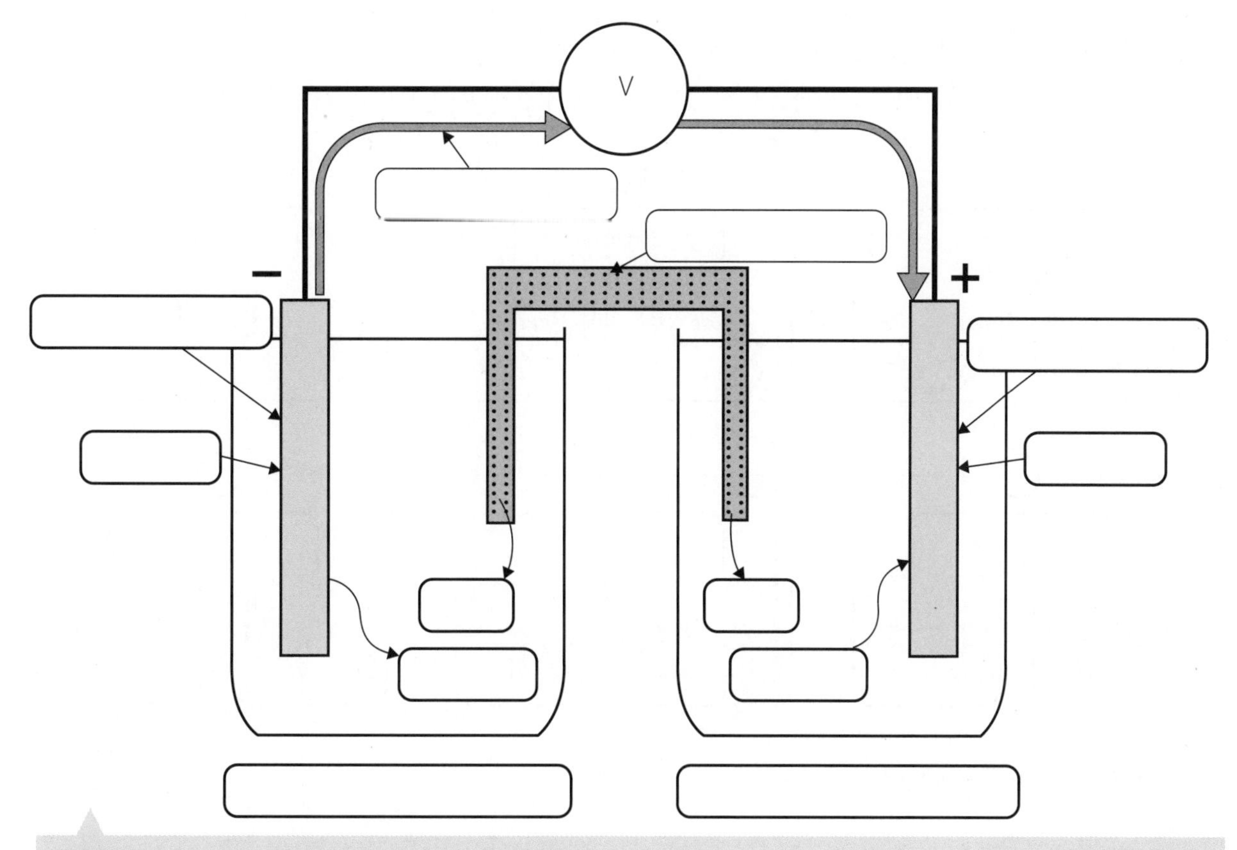

FIGURE 11.1.1 A galvanic cell using nickel and lead electrodes

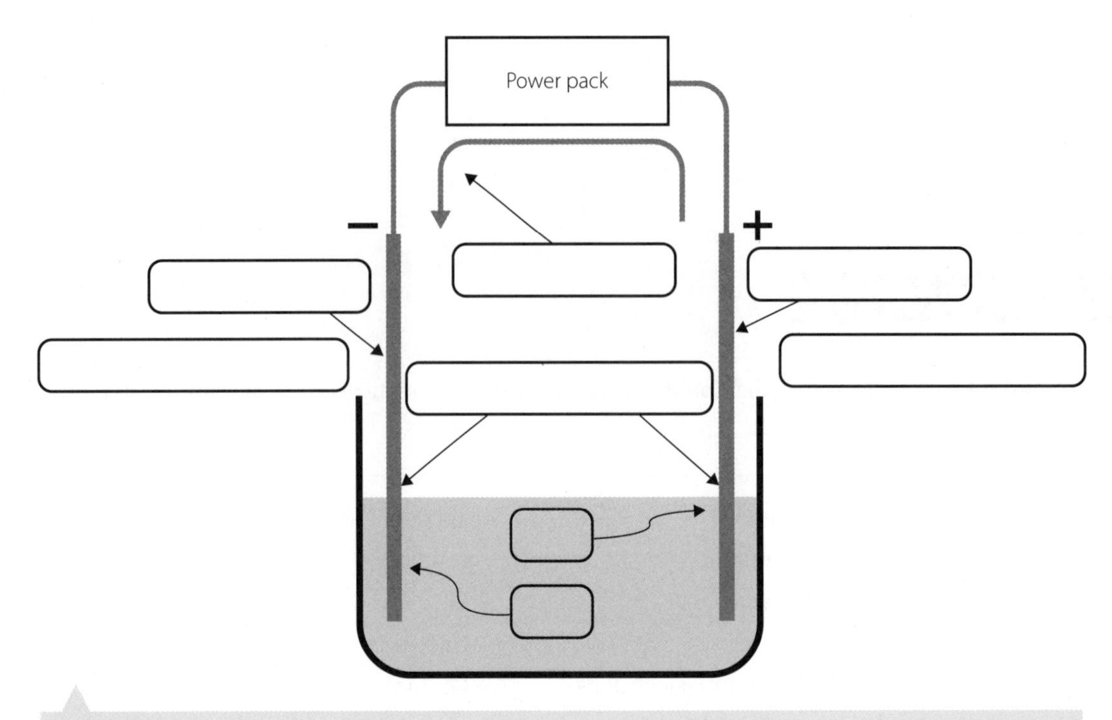

FIGURE 11.1.2 A molten lithium bromide electrolytic cell

2 a Use the galvanic cells shown in Figure 11.1.3 to complete Table 11.1.1 on page 89. In each case, you should choose the most suitable pair of metals. Each pair of metals can only be used once. The activity series of metals is:

K > Na > Li > Ba > Ca > Mg > Al > Zn > Fe > Ni > Sn > Pb > (H) > Cu > Ag > Pt > Au

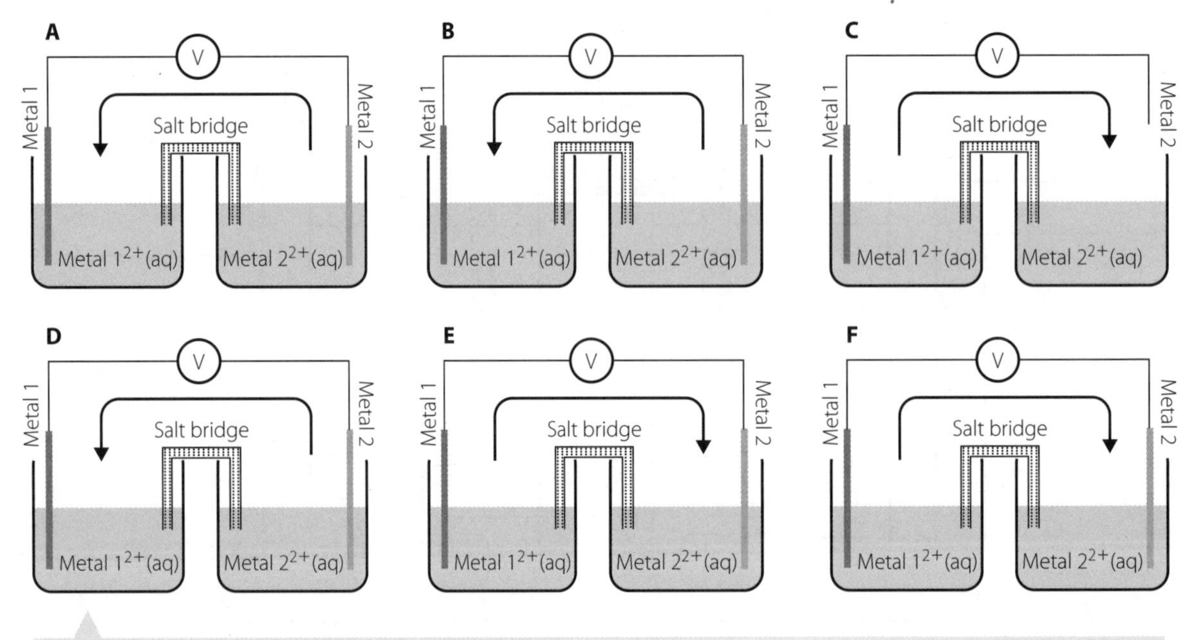

FIGURE 11.1.3 Series of galvanic cells

TABLE 11.1.1 Identification of galvanic cells from Figure 11.1.3

CELL	METAL 1	METAL 2
	Cu	Zn
	Mg	Zn
	Pb	Fe
	Sn	Cu
	Mg	Pb
	Ni	Mg

b Justify your answer for each of the galvanic cells with the aid of half-equations.

3 Complete the table to summarise the differences between galvanic cells and electrolytic cells.

TABLE 11.1.2 Summary of differences between galvanic and electrolytic cells

	GALVANIC CELL	ELECTROLYTIC CELL
Reaction spontaneity		
Power source		
Direction of electron flow		
Anode charge		
Cathode charge		
Location of oxidation and reduction		
Number of cells and why		
Electrode polarity determined by		

9780170412476

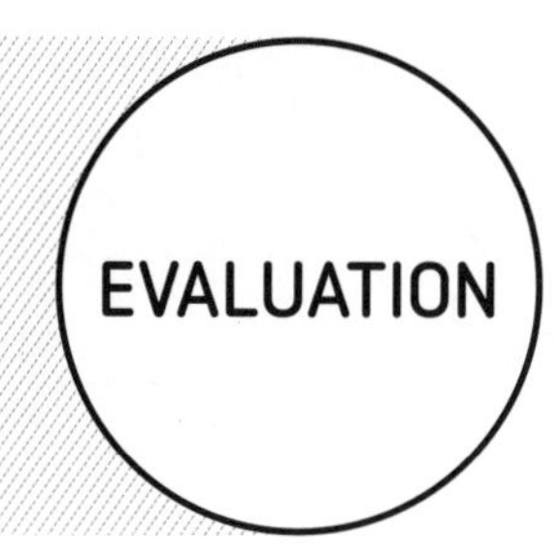

EVALUATION

1 The half-reaction that occurs at the anode during the electrolysis of molten lithium bromide is:

A $Li(l) \rightarrow Li^{+}(l) + e^{-}$.

B $Br_2(l) + 2e^{-} \rightarrow 2Br^{-}(l)$.

C $Li^{+}(l) + e^{-} \rightarrow Li(l)$.

D $2Br^{-}(l) \rightarrow Br_2(l) + 2e$.

2 In a galvanic cell the role of the salt bridge is to:

A move electrons from one electrolyte to the other.

B move positively charged ions between cells.

C maintain the electrical neutrality of each half-cell.

D ensure electrons remain in each beaker.

3 Which of the following equations best represents a galvanic cell using copper and nickel electrodes?

A $Ni^{2+}(aq) + Cu(s) \rightarrow Cu^{2+}(aq) + Ni(s)$

B $Ni(s) + Cu(s) \rightarrow Ni^{2+}(aq) + Cu^{2+}(aq)$

C $Ni^{2+}(aq) + Cu^{2+}(aq) \rightarrow Ni(s) + Cu^{2+}$

D $Cu^{2+}(aq) + Ni(s) \rightarrow Ni^{2+}(aq) + Cu(s)$

4 Electrolytic cells have one cell and galvanic cells have two half-cells because:

A the reaction in an electrolytic cell is spontaneous.

B the reactants in a galvanic cell need to be kept separate because they react spontaneously.

C electrolytic cells have a source of power.

D electrons produced in a galvanic cell can travel through the wires of the external circuit.

5 'In a voltaic cell, the cathode is comprised of the *reduced species*.' With the aid of half-equations examine the accuracy of this statement.

6 Write the half-equations representing the electrolysis of molten magnesium chloride.

7 Draw and label a galvanic cell represented by the equation:

$$Ni(s) + 2AgNO_3(aq) \rightarrow Ni(NO_3)_2(aq) + 2Ag(s)$$

12 Galvanic cells

Summary

- A galvanic cell is an electrochemical cell in which the oxidation and reduction chemical reactions occur spontaneously.
- The electrons produced by oxidation at the anode move through the external circuit to reduce the species at the cathode.
- The simplest galvanic cells can be represented by spontaneous metal displacement reactions in which electrons are transferred from a metallic reducing agent to a metallic ion oxidising agent.
- Due to the spontaneity of these reactions, the reacting species have to be separated into separate containers (half-cells). The electrons are forced to travel through the wires of the external circuit, where they can be harnessed.
- In order for a galvanic cell to operate, a salt bridge is used to complete the circuit. A salt bridge is necessary to maintain the electrical neutrality of each half-cell.

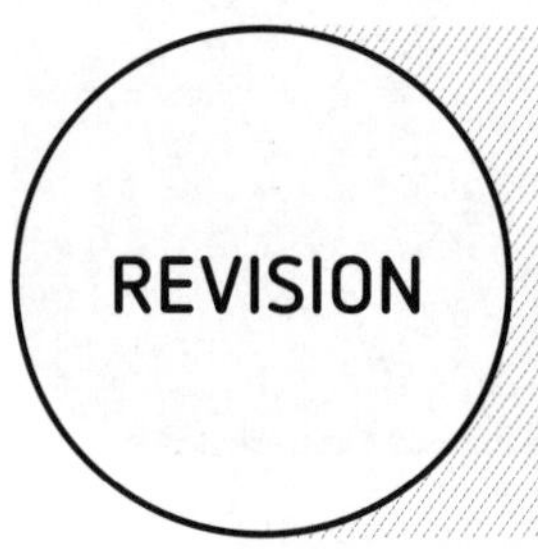

12.1 Galvanic cells

Galvanic cells can be constructed using metal displacement reactions.

For example, the reaction between zinc metal and copper(II) sulfate solution:

$$Zn(s) + CuSO_4(aq) \rightarrow ZnSO_4(aq) + Cu(s)$$

Writing the net ionic equation gives:

$$Zn(s) + Cu^{2+}(aq) \rightarrow Zn^{2+}(aq) + Cu(s)$$

Separating the reactants and writing their half-equations gives:

$$Zn(s) \rightarrow Zn^{2+}(aq) + 2e^- \quad \textit{Oxidation}$$

$$Cu^{2+}(aq) + 2e^- \rightarrow Cu(s) \quad \textit{Reduction}$$

Each half-equation is the basis of a half-cell.

The *oxidation* half-cell consists of a beaker containing zinc sulfate solution with a strip of zinc metal dipping into it.

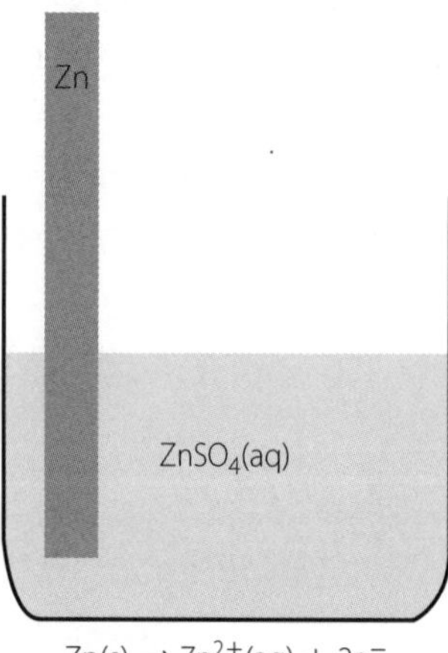

The *reduction* half-cell consists of a beaker containing copper(II) sulfate solution with a strip of copper metal dipping into it.

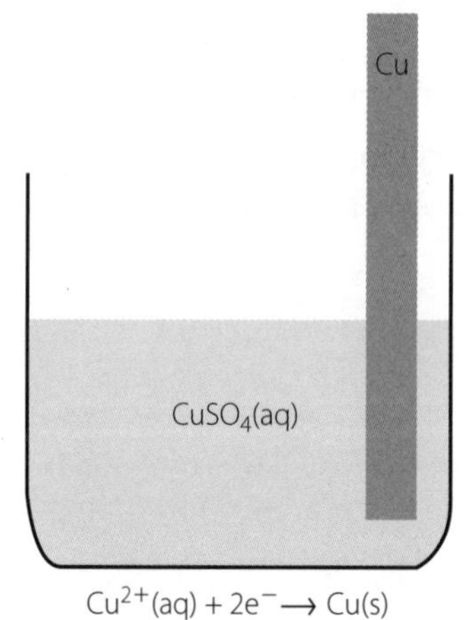

FIGURE 12.1.1 The oxidation and reduction half-cells

9780170412476

The set-up for connecting the two half-cells to an external circuit consisting of wires and a voltmeter is shown in Figure 12.1.2.

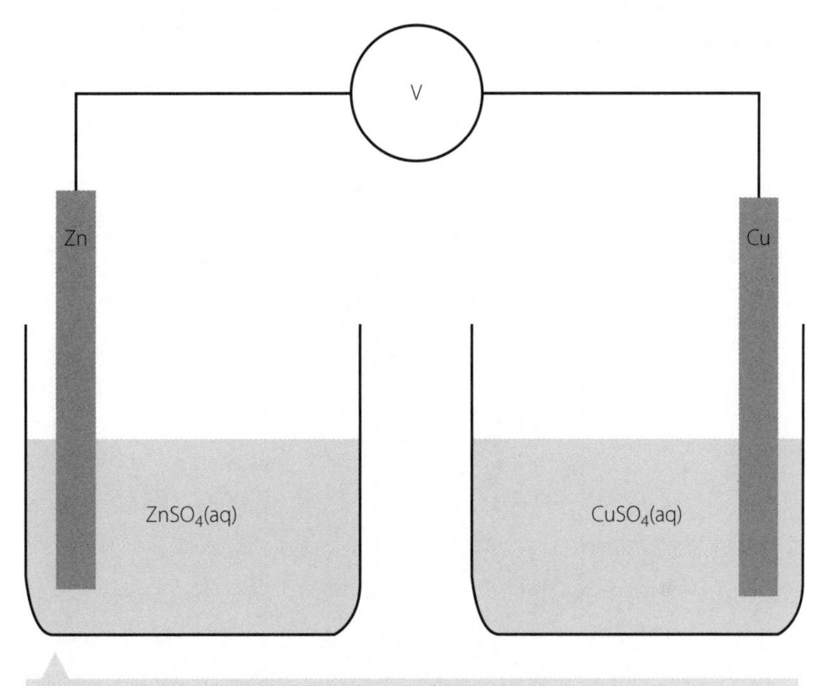

FIGURE 12.1.2 The half-cells connected to an external circuit

The circuit is completed by the addition of a salt bridge.

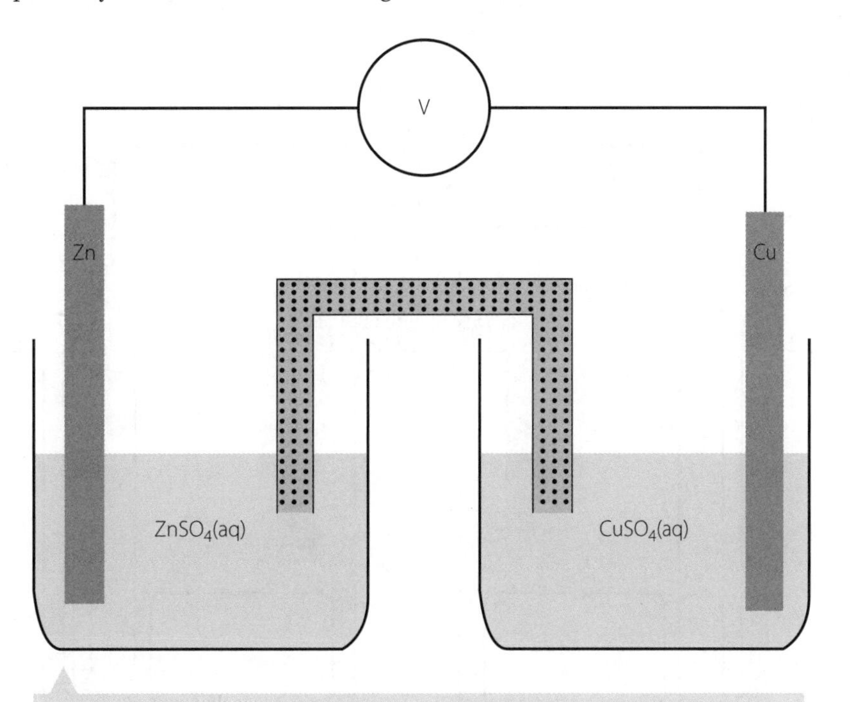

FIGURE 12.1.3 Addition of a salt bridge completes the galvanic cell

1 The figures below show a diagram of an alkali hydrogen fuel cell (Figure 12.1.4) and a generalised galvanic cell (Figure 12.1.5). An alkali fuel cell is simply a compartmentalised galvanic cell. Use the information given for the alkali fuel cell in Figure 12.1.4 to complete the diagram of the generalised galvanic cell in Figure 12.1.5. Make sure to include the direction of electron flow.

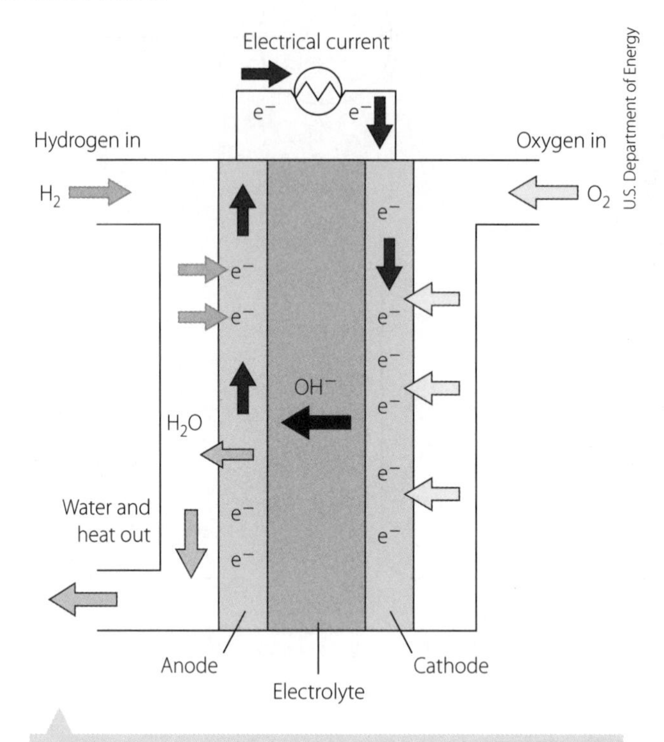

FIGURE 12.1.4 Alkali fuel cell

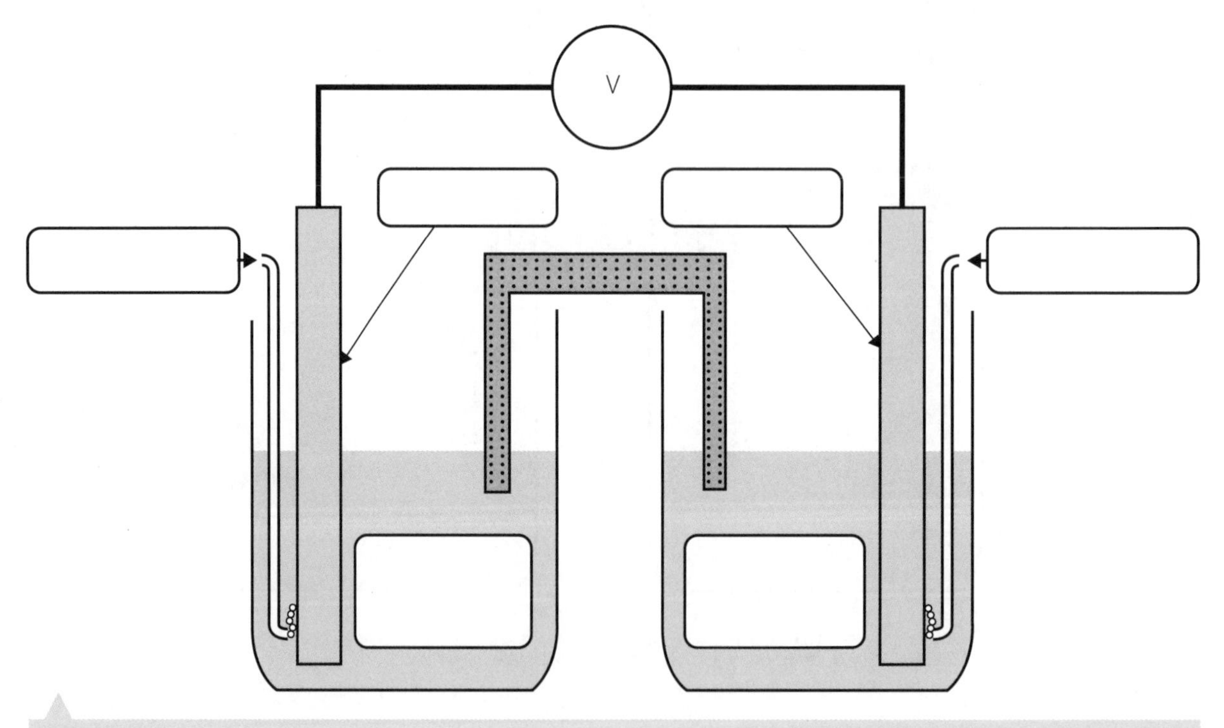

FIGURE 12.1.5 Galvanic cell

2 An experiment was carried out in which a number of galvanic cells were constructed using zinc, iron, copper and lead electrodes dipped into solutions of their respective ions. The external circuit in each case included a centre-reading galvanometer that measures the voltage produced by each cell and the direction of current.

Figure 12.1.6 shows partial cell diagrams in which only the electrodes and the external circuit are visible. The activity series provides additional information:

K > Na > Li > Ba > Ca > Mg > Al > Zn > Fe > Sn > Pb > H > Cu > Ag > Pt > Au

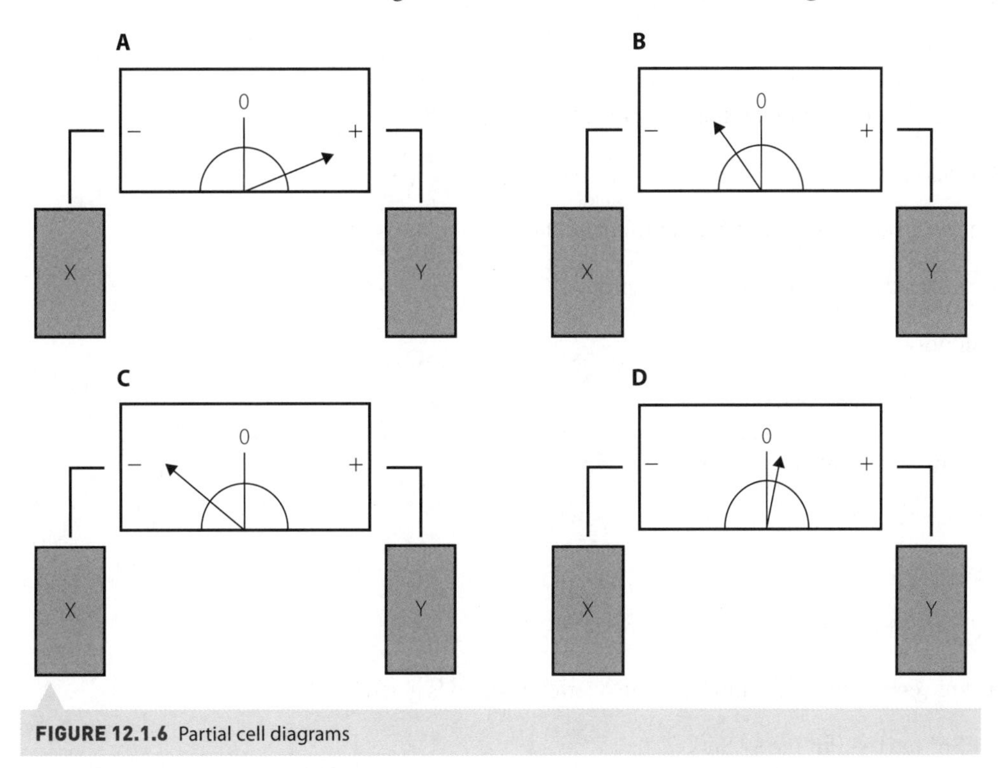

FIGURE 12.1.6 Partial cell diagrams

For each cell, **A–D**, identify the electrodes X and Y by filling in the table with the appropriate metal and writing the oxidation and reduction equations.

	ELECTRODE X	ELECTRODE Y
A		
B		
C		
D		

EVALUATION

1 In a galvanic cell, which half-reaction takes place at the anode?

A The reduction

B The one that transfers the greatest number of electrons

C The one with the greater reduction potential

D The oxidation

2 In a voltaic cell, the mass of the:

A anode stays the same.

B cathode decreases.

C anode decreases.

D cathode stays the same.

3 In the galvanic cell between tin and lead, the overall reaction is given by:

A $Pb(s) + Sn^{2+}(aq) \rightarrow Pb^{2+}(aq) + Sn(s)$

B $Pb^{2+}(aq) + Sn^{2+}(aq) \rightarrow Pb(s) + Sn(s)$

C $Pb^{2+}(aq) + Sn(s) \rightarrow Sn^{2+}(aq) + Pb(s)$

D $Pb(s) + Sn(s) \rightarrow Pb^{2+}(aq) + Sn^{2+}(aq)$

4 In an alkali fuel cell:

A hydrogen gas is the positive electrode.

B the electrolyte is water.

C electrons flow from the oxygen electrode to the hydrogen electrode.

D oxygen gas is the cathode.

5 Explain why, in a galvanic cell, the salt bridge usually contains a nitrate salt of sodium or potassium.

---

6 During the operation of a galvanic cell using tin and copper electrodes, explain why the cathode gains mass.

9780170412476

7 With the aid of half-equations, explain why an alkali fuel cell needs to have both electrodes lined with platinum.

8 Referring to the graph below, explain why the voltage produced by the galvanic cell decreases over time.

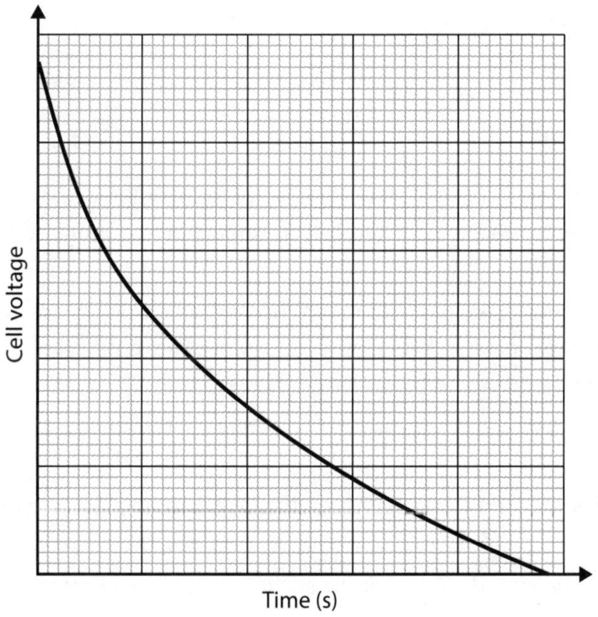

FIGURE 12.2.1 Graph of cell voltage against time for a galvanic cell

13 Standard electrode potential

Summary

- The standard electrode potential, $E°$, of a substance is the potential difference generated by a galvanic cell when the substance making up a half-cell is connected to a standard hydrogen half-cell.
- When a hydrogen half-cell is connected to the standard hydrogen half-cell the value of $E°$ is 0 V. This is self-evident: the cells are identical. When other substances are connected to the standard hydrogen half-cell they generate positive or negative $E°$ values.
- The sign of the $E°$ value of a substance is a measure of the ability of the substance to gain or lose electrons relative to hydrogen.
- A positive $E°$ value for a substance indicates that the substance is more likely to gain electrons than hydrogen.
- A negative $E°$ value for a substance indicates that the substance is more likely to lose electrons than hydrogen.
- When the standard electrode potentials of substances making up a half-cell connected to the hydrogen half-cell are measured, an electrochemical series is produced that is invaluable when predicting cell potentials and oxidising and reducing agents.
- In an electrochemical series, the strongest oxidising agents have the greatest positive $E°$ values while the strongest reducing agents have the greatest negative values.
- Standard electrode potentials can be combined to determine cell potentials between two substances. The sign of the cell potential indicates whether a reaction will be spontaneous and which substance will be the anode in a galvanic cell.

REVISION

13.1 Electrochemical series

1 The following table shows part of the electrochemical series.

	Oxidised form ne^- → Reduced form	$E°$(V)
[ ]	$F_2 + 2e^- \rightarrow 2F^-$	__2.87
	$Au^+ + e^- \rightarrow Au$	__1.69
	$Cl_2 + 2e^- \rightarrow 2Cl^-$	__1.36
	$Br_2 + 2e^- \rightarrow 2Br^-$	__1.09
	$Cu^{2+} + 2e^- \rightarrow Cu$	__0.34
	$2H^+ + 2e^- \rightarrow H_2$	0.00
	$Pb^{2+} + 2e^- \rightarrow Pb$	__0.13
	$Ni^{2+} + 2e^- \rightarrow Ni$	__0.26
	$Cd^{2+} + 2e^- \rightarrow Cd$	__0.40
	$Fe^{2+} + 2e^- \rightarrow Fe$	__0.45
	$Zn^{2+} + 2e^- \rightarrow Zn$	__0.76
	$Mg^{2+} + 2e^- \rightarrow Mg$	__2.37
	$Na^+ + e^- \rightarrow Na$	__2.71
[ ]	$K^+ + e^- \rightarrow K$	__2.93

a On the table, add a + or – to the $E°$ values for each half-equation.

b Decide, on the basis of their respective $E°$ values, which element is the strongest oxidising agent and which is the strongest reducing agent. Put these descriptions in the box provided next to the appropriate elements.

c Suggest why all of the half-equations in this table are reduction half-equations.

____

____

____

____

2 A number of experiments were carried out in which strips of unknown metals were systematically dipped into solutions of the metal ions. The results are summarised in Table 13.1.1.

TABLE 13.1.1 Summary of results for some metal displacement experiments

	$L^{2+}(aq)$	$Q^{2+}(aq)$	$T^{2+}(aq)$	$X^{2+}(aq)$	$R^{2+}(aq)$	$G^{2+}(aq)$
L		✓	✗	✓	✓	✗
Q	✗		✗	✗	✗	✗
T	✓	✓		✓	✓	✓
X	✗	✓	✗		✓	✗
R	✗	✓	✗	✗		✗
G	✓	✓	✗	✓	✓	

Using the results from Table 13.1.1, arrange the elements according to their reactivity and place them in the appropriate position in Table 13.1.2, along with their reduction half-equation.

TABLE 13.1.2 Summary of elements according to their reactivity

Oxidised form $2e^-$ → Reduced form	$E°$ (V)
	−0.32
	+0.78
$T^{2+}(aq) + 2e^- \rightarrow T(s)$	−0.80
	−0.08
	+0.29
	−0.42

9780170412476

13.2 Determining standard cell potentials

When a metal displacement reaction is used in a galvanic cell, it is possible to use each metal's standard electrode potential to determine the overall cell voltage.

WORKED EXAMPLE

Predict the cell potential of a galvanic cell constructed using manganese and copper electrodes.

ANSWER

a Refer to a table of standard electrode potentials to identify the metals, their reduction half-equations and their standard electrode potentials.

$$Mn^{2+} + 2e^- \rightarrow Mn \qquad E° = -1.18\,V$$

$$Cu^{2+} + 2e^- \rightarrow Cu \qquad E° = +0.34\,V$$

b Determine which of the metals is the strongest oxidising agent and which metal is the strongest reducing agent. This determines which will be the anode and which will be the cathode.

The Mn half-equation has a much smaller electrode potential than Cu.
It loses its electrons (is oxidised) much more easily than Cu.
Mn is therefore the anode and Cu is the cathode.

c Use the equation:

$$E°\text{cell} = \underbrace{E°\text{red} + E°\text{ox}}_{\text{REDOX}}$$

to determine the cell potential.

Remember: The Mn is being oxidised so its reduction half-equation must be reversed. If the half-equation is reversed, the sign of its $E°$ value must also be reversed.

Therefore:

$$Mn^{2+} + 2e^- \rightarrow Mn,\ E° = -1.18\,V$$

becomes:

$$Mn \rightarrow Mn^{2+} + 2e^-,\ E° = +1.18\,V$$

$$E° = +0.34 + (+1.18) = +1.52\,V$$

QUESTION

1 A student set up a number of galvanic cells, attaching a digital voltmeter to each one. Figure 13.2.1 shows the electrodes of each cell connected to the digital voltmeter.

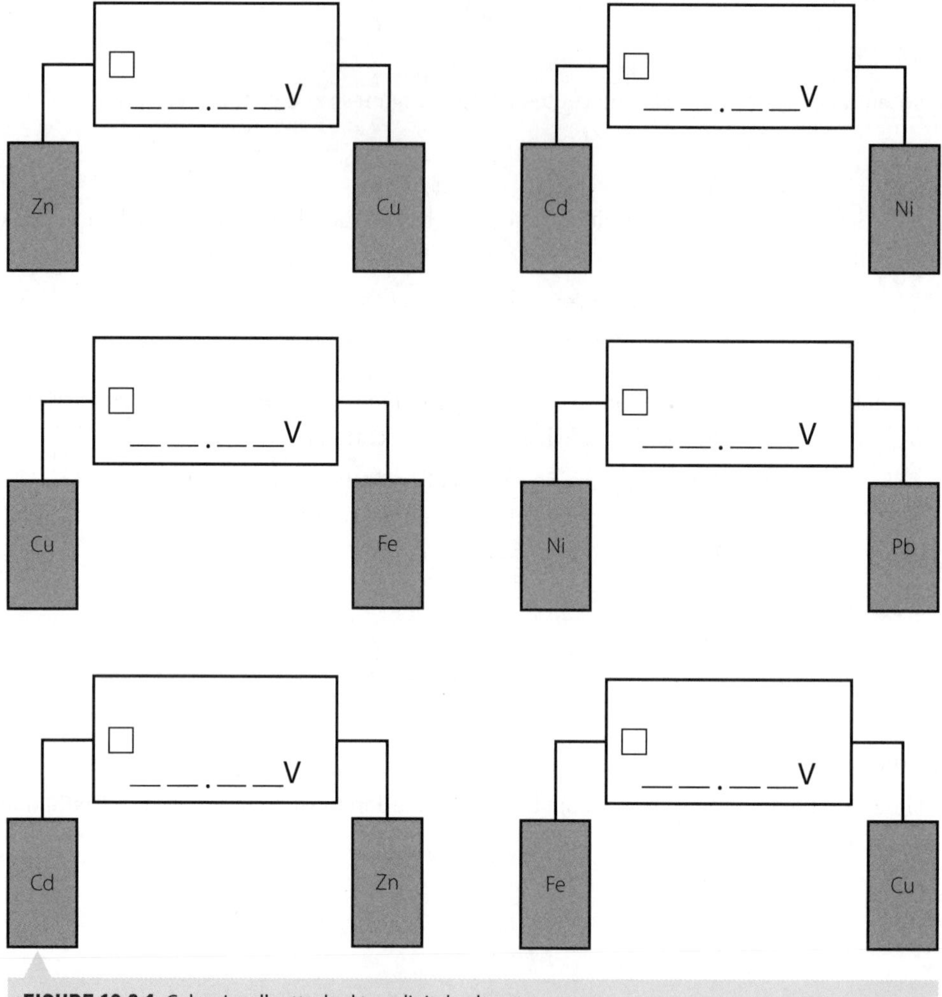

FIGURE 13.2.1 Galvanic cells attached to a digital voltmeter

a Complete the figure by determining what the reading on each voltmeter will be.

b From these results, decide which reactions are non-spontaneous.

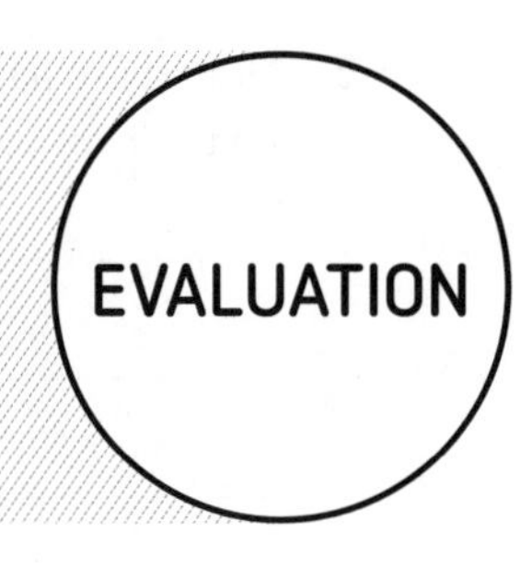

1 Determine the emf of a cell represented by the equation:

$$Cd(s) + 2AgNO_3(aq) \longrightarrow Cd(NO_3)_2(aq) + 2Ag(s)$$

Refer to Table 13.1.1 of standard electrode potentials on page 181 of the textbook.

A −1.2 V

B +2.0 V

C +0.4 V

D +1.2 V

2 The reduction half-equations and standard electrode potentials for two substances, M and N, are given by:

$$M^{2+}(aq) + 2e^- \longrightarrow M(s) \qquad E° = -0.6\,V$$

$$L^+(aq) + e^- \longrightarrow L(s) \qquad E° = +0.73\,V$$

If L(s) and M^+(aq) were mixed, it is clear that:

A M would be the reducing agent.

B L would displace M^{2+} from solution.

C the reaction would be non-spontaneous.

D L would be the oxidising agent.

3 In the galvanic cell using nickel and gold electrodes, the mass of the gold electrode increases. This is because:

A the gold has a more negative standard electrode potential than nickel.

B the nickel has a more negative standard electrode potential than gold.

C the cell emf is negative.

D the gold can displace nickel ions from solution.

4 A galvanic cell consists of a platinum electrode and an unknown substance. The emf of the cell is +2.36 V. What is the other substance?

A Mn

B I_2

C Mg

D Zn

5 Aluminium has a standard electrode potential of −1.66 V. Explain why, when mixed with copper(II) sulfate solution, no observable reaction occurs.

6 Metal Q has a standard electrode potential of −0.62 V, while metal T has a standard electrode potential +0.38 V. Which metal should be used as the cathode in a galvanic cell? With the aid of half-equations, justify your choice.

7 A galvanic cell is constructed using tin and silver electrodes. The cell can be represented by the equation:

$$Sn(s) + 2Ag^{+}(aq) \rightarrow Sn^{2+}(aq) + 2Ag(s)$$

Explain why, when the emf is calculated, the $E°$ value for silver is not multiplied by 2 to reflect the mole ratio in the balanced equation.

9780170412476

14 Electrolytic cells

Summary

- In electrolytic cells, an external power source such as a battery or power pack is used to force a non-spontaneous reaction to occur.
  For example, the decomposition of sodium chloride into sodium metal and chlorine gas would never occur spontaneously. However, passing an electric current through molten sodium chloride forces this reaction to occur.
- In an electrolytic cell, oxidation occurs at the anode.
- When molten sodium chloride is electrolysed, the half-equations are:

$$Na^+(l) + e^- \rightarrow Na(l)$$

$$2Cl^-(l) \rightarrow Cl_2(g) + 2e^-$$

- As with galvanic cells, the standard electrode potentials of the reactants can be used to determine the potential of the electrolytic cell.
- Cell potential of electrolytic cells is *always* positive. A positive cell potential indicates that the reaction is non-spontaneous.
- When aqueous salt solutions are electrolysed the presence of water needs to be considered.
- When predicting the products of the electrolysis of aqueous solutions at the anode and the cathode, the half-cell reaction with the lowest $E°$ value is the one that occurs.
- The concentration of the electrolyte can affect the products of electrolysis.
- There are a number of industrially important applications of electrolysis including electroplating, electrorefining of metals and extraction of important elements.

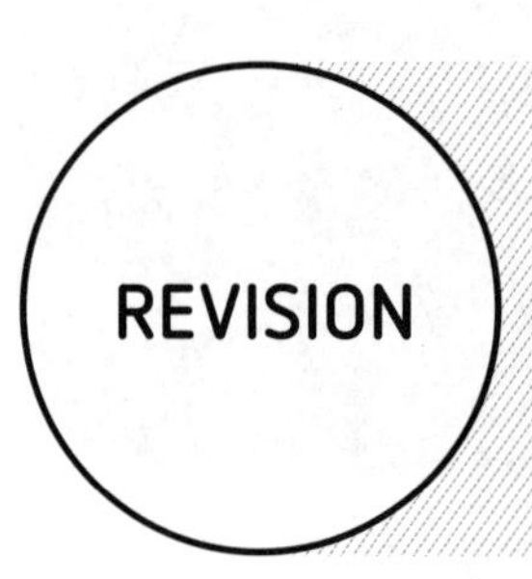

14.1 Electrolysis of molten salts

The electrolysis of simple salts, such as sodium chloride, involves melting the salt so that the ions can move around freely, which allows for the conduction of electrical current.

The reaction:

$$2NaCl(l) \rightarrow 2Na(l) + Cl_2(g)$$

is not spontaneous. Table salt does not spontaneously decompose to highly reactive sodium metal and deadly poisonous chlorine gas!

Splitting the reaction into its half-equations gives:

$$2Na^+(l) + 2e^- \rightarrow Na(l) \quad \textit{Reduction}$$

$$2Cl^-(l) \rightarrow Cl_2(g) + 2e^- \quad \textit{Oxidation}$$

This can be confirmed by calculating the emf of the reaction:

$$Na^+ + e^- \rightarrow Na \qquad E° = -2.71\,V$$

$$Cl_2 + 2e^- \rightarrow 2Cl^- \quad E° = +1.36\,V$$

(This must be reversed because the Cl^- ions are oxidised.)

$$E°_{cell} = E°_{red} + E°_{ox}$$

$$= -2.71 + (-1.36)$$

$$= -4.07\,V$$

The negative sign indicates that this is a non-spontaneous reaction.

For the electrolysis of molten sodium chloride, at least 4.07 V of electricity must be supplied. Figure 14.1.1 shows a diagram of the electrolysis of sodium chloride.

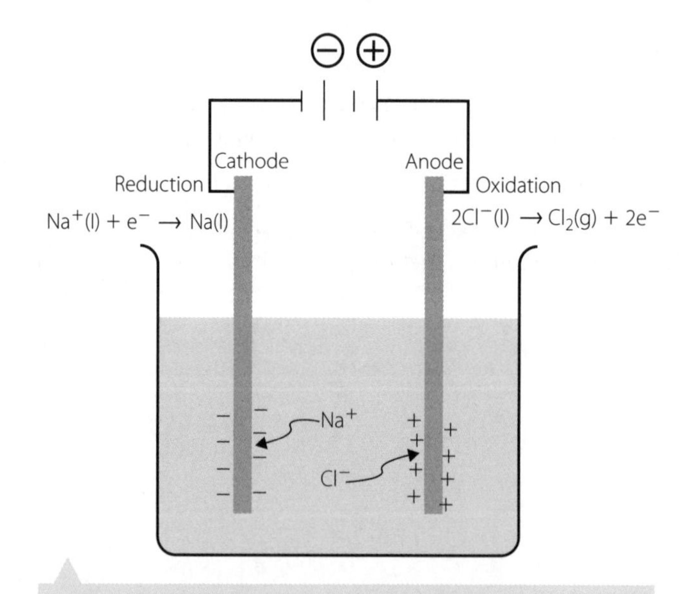

FIGURE 14.1.1 The electrolysis of molten sodium chloride

9780170412476

This is the basic process behind the industrial production of sodium metal.

1 The industrial production of sodium takes place in a Downs cell, as shown in Figure 14.1.2.

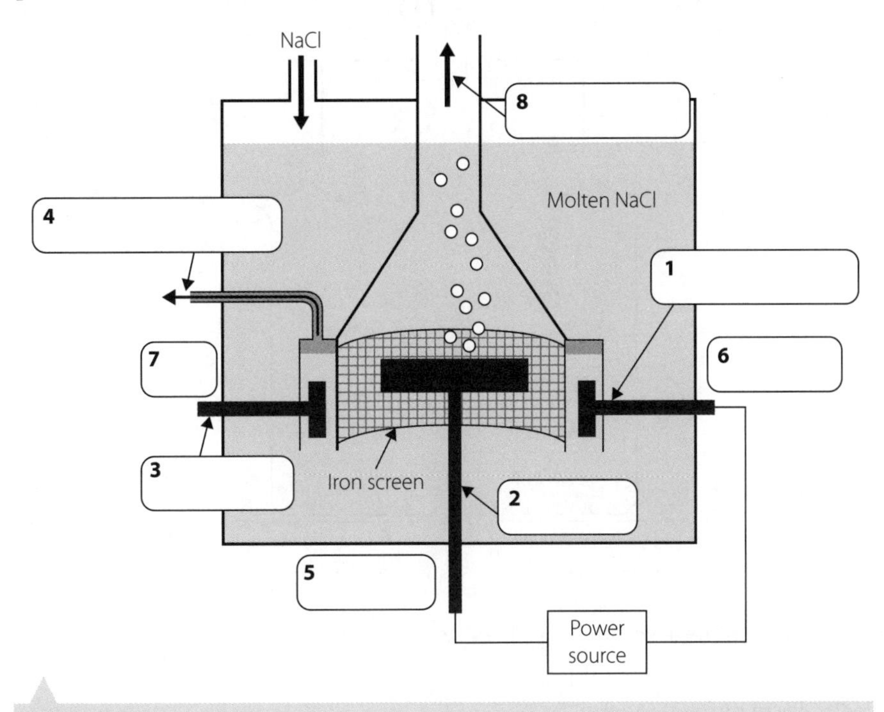

FIGURE 14.1.2 Diagrammatic representation of a Downs cell for the industrial production of sodium

This is a large-scale adaptation of the simple cell shown in Figure 14.1.1.
Use your understanding of the electrolysis of molten salts to add appropriate labels to the boxes in the figure (numbered **1**–**8**). Ensure to include the charge at each electrode.

14.2 Electrolysis of aqueous solutions

An aqueous solution can be defined as a salt dissolved in water. The presence of water complicates the process, because in addition to the electrolysis of the ions of the electrolyte, the water itself can be electrolysed. The half-equations for the electrolysis of water are:

$$\textit{Oxidation}: 2H_2O(l) \rightarrow O_2(g) + 4H^+(aq) + 4e^- \qquad E° = -1.23\,V$$

$$\textit{Reduction}: 2H_2O(l) + 2e^- \rightarrow H_2(g) + 2OH^-(aq) \qquad E° = -0.83\,V$$

When deciding which products are formed at the anode and cathode, remember,

the reaction that is more likely to occur is the one that has the least negative voltage.

1 Consider the electrolysis of tin(II) bromide, $SnBr_2(aq)$, using the apparatus shown in Figure 14.2.1.

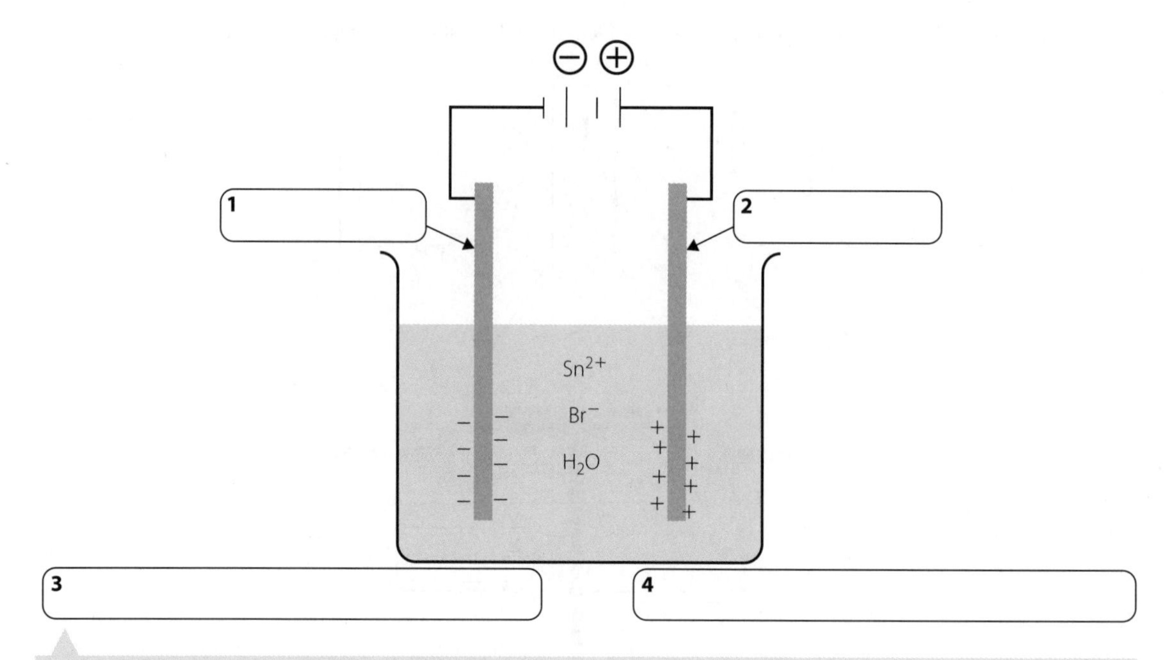

FIGURE 14.2.1 Electrolysis of tin(II) bromide

a Label the electrodes in boxes **1** and **2**.

b Boxes **3** and **4** refer to the preferred reaction at each electrode. Examine the half-equations in Table 14.2.1 and decide which one will be preferred at each electrode.

TABLE 14.2.1 Half-equations and standard electrode potentials

HALF-EQUATION	$E°$ (V)
$Sn^{2+}(aq) + 2e^- \rightarrow Sn(s)$	−0.14
$2Br^-(aq) \rightarrow Br_2(l) + 2e^-$	−1.09
$2H_2O(l) + 2e^- \rightarrow H_2(g) + 2OH^-(aq)$	−0.83
$2H_2O(l) \rightarrow O_2(g) + 4H^+(aq) + 4e^-$	−1.23

ELECTROLYSIS OF AQUEOUS SOLUTIONS: COMPLICATIONS

When deciding upon the products of the electrolysis of aqueous solutions, two further complications need to be considered:

- the nature of the electrolyte
- the concentration of the electrolyte.

The nature of the electrolyte

Some ions will never be electrolysed to their element in aqueous solutions.

For example, positively charged ions with a standard electrode potential of greater negative potential than the reduction of water cannot be electrolysed.

TABLE 14.2.2 Half-equations and standard electrode potentials

<table>
<tr><th></th><th>HALF-EQUATION</th><th>$E°$ (V)</th></tr>
<tr><td></td><td>$2H_2O(l) + 2e^- \rightarrow H_2(g) + 2OH^-(aq)$</td><td>−0.83</td></tr>
<tr><td></td><td>$Mn^{2+}(aq) + 2e^- \rightarrow Mn(s)$</td><td>−1.18</td></tr>
<tr><td rowspan="5">Will not be reduced in aqueous solution</td><td>$Al^{3+}(aq) + 3e^- \rightarrow Al(s)$</td><td>−1.66</td></tr>
<tr><td>$Mg^{2+}(aq) + 2e^- \rightarrow Mg(s)$</td><td>−2.37</td></tr>
<tr><td>$Na^+(aq) + e^- \rightarrow Na(s)$</td><td>−2.71</td></tr>
<tr><td>$Ca^{2+}(aq) + 2e^- \rightarrow Ca(s)$</td><td>−2.87</td></tr>
<tr><td>$Ba^{2+}(aq) + 2e^- \rightarrow Ba(s)$</td><td>−2.91</td></tr>
<tr><td></td><td>$K^+(aq) + e^- \rightarrow K(s)$</td><td>−2.93</td></tr>
<tr><td></td><td>$Li^+(aq) + e^- \rightarrow Li(s)$</td><td>−3.04</td></tr>
</table>

Polyatomic ions are very stable and would require large voltages to be oxidised. When polyatomic ions are involved, the water will always be oxidised in preference.

Concentration of the electrolyte

When there are competing reactions at an electrode, the reaction with the least negative voltage is the one that occurs.

However, if the $E°$ values of the competing reactions are very similar, then concentration of the electrolyte can have an effect.

The standard electrode potentials given are for 1 M solutions.

2 A number of experiments were carried out in which various salt solutions were electrolysed.

TABLE 14.2.3 Results of experiments

EXPERIMENT	SOLUTION	CONCENTRATION (M)
1	$NaCl$	8.00
2	$CuSO_4$	1.00
3	$Zn(NO_3)_2$	1.00
4	$CuCl_2$	1.00
5	$ZnCl_2$	0.001

A few drops of universal indicator were placed in each cell. In acidic solution, universal indicator turns red. In basic solution, universal indicator turns blue.

Observations were made for each experiment and are summarised in Table 14.2.4. However, the experiment number was not recorded in each case. Each experiment was arbitrarily assigned a letter.

In some cases, a gas was produced at one or both of the electrodes. Identification of these gases is achieved using the following tests/observations.

- *Hydrogen gas:* colourless, odourless. If a lit splint or match is placed in a test tube containing hydrogen, the gas combusts, producing a 'pop' sound.
- *Oxygen gas:* colourless, odourless. If a glowing splint or match is placed in a test tube containing oxygen, the splint bursts back into flame.
- *Chlorine gas:* greenish-colour, sharp-smelling. If moist blue litmus paper is placed in the mouth of test tube containing chlorine, the litmus paper is bleached and turns white.

TABLE 14.2.4

EXPERIMENT	ELECTRODE X	ELECTRODE Y
A	Greenish-coloured gas produced. Bleaches moist blue litmus paper.	Red–brown deposit appeared on the electrode.
B	Colourless, odourless gas produced. Relights glowing splint. Solution turned orange-pink around electrode.	Red–brown deposit appeared on the electrode.
C	Colourless, odourless gas produced. Relights glowing splint. Solution turned orange-pink around electrode.	Colourless, odourless gas produced. Successfully underwent 'pop' test. Solution turned blue around electrode.
D	Greenish-coloured gas produced. Bleaches moist blue litmus paper.	Colourless, odourless gas produced. Successfully underwent 'pop' test. Solution turned blue around electrode.
E	Colourless, odourless gas produced. Relights glowing splint. Solution turned orange-pink around electrode.	Silvery deposit appeared on the electrode.

a Identify electrodes **X** and **Y**.

____

b In the table below, match the experiment letter with the experiment number and add the appropriate half-equation in each case.

LETTER	NUMBER	ANODE HALF-EQUATION	CATHODE HALF-EQUATION
A			
B			
C			
D			
E			

c For each experiment, calculate the minimum voltage required to power the cell.

LETTER	NUMBER	CALCULATION	VOLTAGE REQUIRED
A		$E^\circ_{cell} = E^\circ_{red}$ ______ $+ E^\circ_{ox}$ ______ = ______	
B		$E^\circ_{cell} = E^\circ_{red}$ ______ $+ E^\circ_{ox}$ ______ = ______	
C		$E^\circ_{cell} = E^\circ_{red}$ ______ $+ E^\circ_{ox}$ ______ = ______	
D		$E^\circ_{cell} = E^\circ_{red}$ ______ $+ E^\circ_{ox}$ ______ = ______	
E		$E^\circ_{cell} = E^\circ_{red}$ ______ $+ E^\circ_{ox}$ ______ = ______	

9780170412476

1 In an electrolytic cell the cathode is:

A negative, where oxidation occurs.

B positive, where reduction occurs.

C positive, where oxidation occurs.

D negative, where reduction occurs.

2 In the electrolysis of an 8 M zinc bromide solution, the product at the anode would be:

A oxygen gas.

B zinc metal.

C bromine liquid.

D hydrogen gas.

3 In an electrolytic cell used for the electrorefining of copper, a piece of impure copper is made the:

A anode so that copper metal is oxidised.

B cathode so that copper ions are attracted to it and are reduced.

C anode so that copper ions are attracted to it and are reduced.

D cathode so that copper metal is oxidised.

4 What is the minimum voltage required to operate a cell used for the electrolysis of a 1 M solution of magnesium bromide?
Refer to Table 13.1.1 of standard electrode potentials on page 181 of the textbook.

A 3.94 V

B 1.92 V

C 1.62 V

D 2.06 V

5 Explain why the electrolysis of an 8 M solution of sodium chloride would produce chlorine gas at the anode.

6 In the Downs cell for the industrial production of sodium chloride (section 14.1, question 1), identify an aspect of this process that can be commercially viable, aside from the production of sodium.

7 The alkali fuel cell was described in Chapter 12. This is a galvanic cell in which hydrogen gas and oxygen gas are brought together to produce a voltage of 1.23 V. It could be argued that this cell could be used to power a cell used to electrolyse a dilute sodium chloride solution, thereby producing hydrogen gas and oxygen gas. This could then be used in the alkali fuel cell, creating a closed loop of energy transformations.
With the aid of half-equations, explain why this idea might not work.

8 Complete the figure of an electrolytic cell.

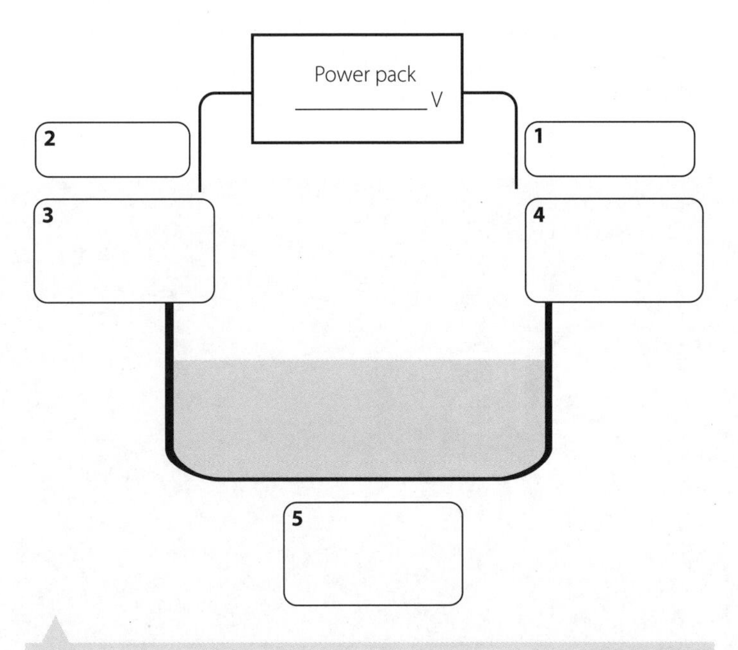

FIGURE 14.3.1 Electrolytic cell used for silver plating a steel bracelet

a Add the electrodes.

b Label boxes **1–5**.

c Calculate the minimum voltage required to operate the cell and write this in the power pack box.

____

____

____

____

____

____

UNIT FOUR

STRUCTURE, SYNTHESIS AND DESIGN

- Topic 1: Properties and structure of organic materials
- Topic 2: Chemical synthesis and design

15 Structure of organic compounds

LEARNING

Summary

- Organic compounds are carbon-based compounds, with a 'skeleton' of carbon atoms attached to atoms of other elements.
- Organic compounds are grouped into homologous series: compounds with the same functional group of atoms and, therefore, similar chemical properties, but varying physical properties according to the length of the carbon chain.
- The structure of an organic molecule can be derived from its name. The name consists of the stem and the prefix of suffix to the stem.
    - The stem indicates the length of the longest carbon chain in the molecule, as shown in Table 15.1.1.

TABLE 15.1.1 The names of the first 10 alkanes

NUMBER OF CARBON ATOMS	NAME STEM	ALKANE
1	Meth-	Methane (CH_4)
2	Eth-	Ethane (C_2H_4)
3	Prop-	Propane (C_3H_8)
4	But-	Butane (C_4H_{10})
5	Pent-	Pentane (C_5H_{12})
6	Hex-	Hexane (C_6H_{14})
7	Hept-	Heptane (C_7H_{16})
8	Oct-	Octane (C_8H_{18})
9	Non-	Nonane (C_9H_{20})
10	Dec-	Decane ($C_{10}H_{22}$)

    - The prefix or suffix to the stem indicates the functional group, the group of atoms which provides the molecule with its distinctive chemical properties, as shown in Table 15.1.2.

TABLE 15.1.2 Some organic compounds and their descriptive features

NAME	FUNCTIONAL GROUP	PREFIX OR SUFFIX	HOMOLOGOUS SERIES	EXAMPLE	NAME OF EXAMPLE
Alkene	—C=C—	-ene	Alkenes	H H C=C H H	Ethene
Alkyne	—C≡C—	-yne	Alkynes	H—C≡C—H	Ethyne
Haloalkane	Halogen —F, —Cl, —Br, —I	Fluoro-, chloro-, bromo-, iodo-	Haloalkanes	H H H—C—C—Br H H	Bromoethane
Hydroxyl	—OH	-anol, or hydroxyl-	Alcohols	H H H—C—C—OH H H	Ethanol
Aldehyde	O ‖ —C—H	-anal	Aldyhydes	H H O H—C—C—C—H H H	Propanal
Ketone	O ‖ —C—	-anone, or oxo-	Ketones	H O H H—C—C—C—H H H	Propanone
Carboxyl	O ‖ —C—OH	-anoic acid	Carboxylic acids	H O H—C—C—OH H	Ethanoic acid
Ester	O ‖ —C—O—R′	alkyl -oate	Esters	O ‖ CH_3—C—O—CH_3	Methyl ethanoate
Amine	—NH_2	-anamine	Amines	H H H H—C—C—N H H H	Ethanamine
Amide	O ‖ —C—NH_2	-anamide	Amides	O H H—C—N H	Methanamide
Nitriles	—C≡N	-nitrile, or cyano-	Nitriles / cyanides	H H—C—C≡N H	Ethane nitrile

- Organic molecules can be represented by a structural or semi-structural formula. A structural formula shows all bonds present. A semi-structural formula lists the carbon atoms in the longest chain in order, with the functional groups next to the atom to which they are attached. Branched chains are shown in parentheses.

For example,

STRUCTURAL FORMULA	SEMI-STRUCTURAL FORMULA
H | H—C—H H	H |	| H—C—C—C—O—H |	| H Cl H	$CH_3CCl(CH_3)CH_2OH$

FIGURE 15.1.1 2-chloro 2-methyl 1-propanol

- Organic molecules are named using a series of conventions.
  1. Number of carbon atoms in the longest continuous chain is established and stem name is derived.
  2. Functional groups identified, named and labelled with the carbon atom to which they are attached. Carbon atoms are numbered in order to keep the numbers as low as possible.
  3. Double or triple bonds are labelled according to the lower number of the two carbon atoms to which they are attached.
  4. Branches in the carbon skeleton are named and numbered in the same way as functional groups.
  5. Prefixes *di-*, *tri-*, or *tetra-* are used to denote more than one of the same functional group or alkyl chain.
  6. Terminal functional groups become carbon atom number one.
  7. Multiple functional groups are listed in alphabetical order.
- Structural isomers are two or molecules with the same molecular formula but differently ordered arrangements of atoms. For example, the molecules in Table 15.1.3 all have the same molecular formula, C_4H_9Cl, and so are all structural isomers of each other.

TABLE 15.1.3 Structural isomers of C_4H_9Cl

STRUCTURAL FORMULA	NAME
H H H H |		| Cl—C—C—C—C—H |		| H H H H	1-chlorobutane
H Cl H H |		| H—C—C—C—C—H |		| H H H H	2-chlorobutane
H H Cl |	| H—C—C—C—H |	| H C H H ⁄	∖ H H	1-chloro 2-methylpropane
H 	 H H—C—H H ∖	⁄ H—C—C—C—H 			 H Cl H	2-chloro 2-methylpropane

- Stereo-isomers are two molecules with a C═C bond that have the same structural formula but different arrangement of atoms in space. The molecules are known as the *cis*- and *trans*-isomers; for example, the two geometrical isometric forms of 2,3-dichloro 2-butene, as shown in Table 15.1.4.

TABLE 15.1.4 Geometric isomers of 2,3-dichloro 2-butene

STRUCTURAL FORMULA	NAME
H H Cl Cl	*cis*-2,3-dichloro 2-butene
H Cl Cl H	*trans*-2,3-dichloro 2-butene

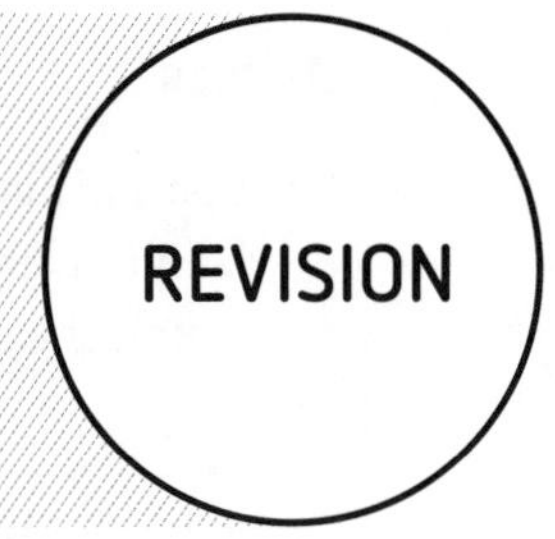

15.1 Important terms

1 Complete the crossword using the clues on page 123.

Across	Down
2 A part of the name of an organic molecule that denotes a functional group and its position on the carbon chain. It comes at the start of the name. **4** __________ formula is the displayed arrangement of the atoms in a molecule, showing all bonds. **6** Carbon ____________ is the arrangement of C atoms in an organic molecule to which other atoms are joined. (8) **8** ___________ series: a series of organic molecules with the same functional group but different length carbon chains **10** Two molecules with the same structural formula but different arrangement of atoms in space are called ____________ isomers. **11** A condensed version of the structural formula showing the sequence of atoms listed by carbon atom but without bonds shown is a ___________ formula.	**1** The ability of carbon atoms to bond with other carbon atoms to form chains and rings **3** __________ group: a group of atoms in a molecule that causes the molecule to chemically react in a distinctive way **5** The part of the name of an organic molecule that denotes the length of the longest carbon chain **7** __________ formula: the number of atoms of each element present in a molecule **9** A part of the name of an organic molecule that denotes a functional group and its position on the carbon chain. It comes at the end of the name.

2 Write a short summary or create a mind map using all of the terms from the crossword.

9780170412476

15.2 Drawing structural and semi-structural formulas

1 Draw structural and semi-structural formulas of the following substances.

a *cis*-pent-2-ene

b 2-butyl bromide

c 2-methyl-1-butanal

d 2-pentanone

e Propanoic acid

f Ethyl propanoate

2 The following compound names have been written incorrectly. Identify what is incorrect with each compound.

a 3-butanol

b 2,2-dichloro-1-pentene

3 Do *cis*- and *trans*-isomers of 2-butene exist? If so, draw structures of their molecules.

4 Do *cis*- and *trans*-isomers of 1, 1-dibromoethene exist? If so, draw structures of their molecules.

15.3 Naming molecules and identifying functional groups

1 The molecular structures of various organic compounds are shown below. For each of the compounds, answer the following questions.

a List any functional group(s) that give rise to the chemical and physical properties of the organic compound.

b What class does the organic compound belong to?

c What is the name of the compound?

d Are there *cis*- and *trans*-isomers of the compound?

Structure	Answers
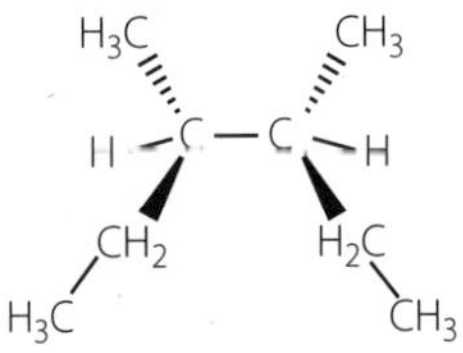 H_3C CH_3 H C — C H CH_2 H_2C H_3C CH_3	**a** **b** **c** **d**
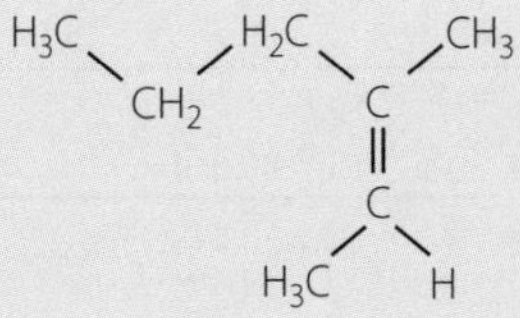 H_3C H_2C CH_3 CH_2 C ‖ C H_3C H	**a** **b** **c** **d**
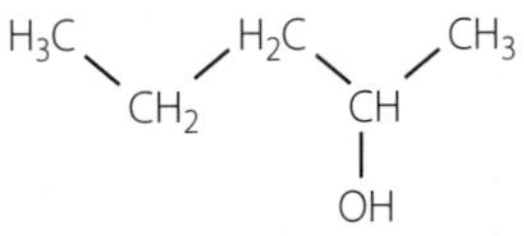 H_3C H_2C CH_3 CH_2 CH OH	**a** **b** **c** **d**
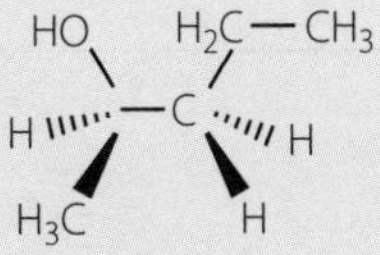 HO H_2C—CH_3 H —C H H_3C H	**a** **b** **c** **d**
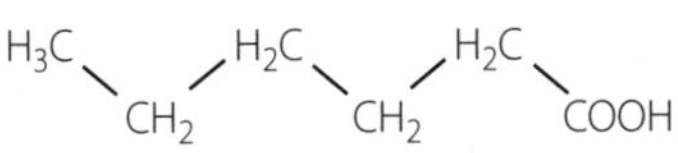 H_3C H_2C H_2C CH_2 CH_2 COOH	**a** **b** **c** **d**

15.4 Geometric isomerism

1 Shown below are four pairs of molecules. Do each of these drawings represent molecules of two different substances that are geometric isomers, or are they identical molecules of one substance? If they represent geometric isomers, which is the *cis*-isomer and which is the *trans*-isomer?

a 1,2-dichloroethane

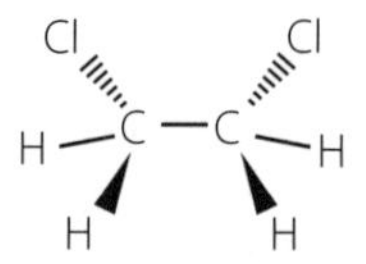

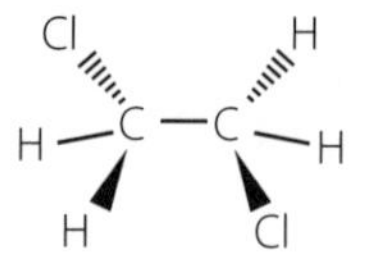

b 2,chloro-1-propane

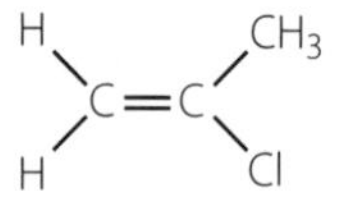

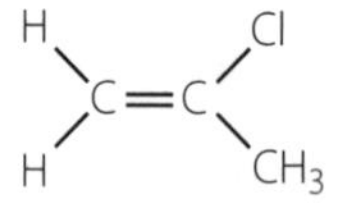

c 1,2-dichloroethene

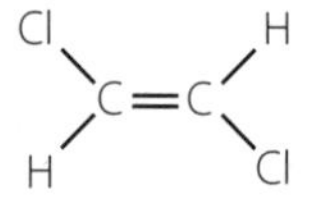

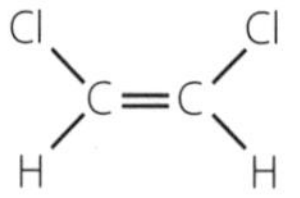

d 1,2-dichloroethene

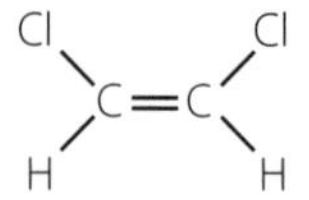

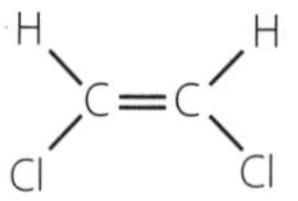

9780170412476

1 The two simplest hydrocarbon groupings are alkanes and alkenes. Both of these groups have some properties in common, but each grouping also has its own unique properties. The groupings are usually referred to as homologous series. The alkane butane has two isomers, whereas the alkene butene has four isomers.

a Define 'homologous series'.

---

b Define 'isomer', and identify and explain the type of isomerism that is displayed by alkenes but not by alkanes.

---

c Draw and name the isomers of butane.

d Draw and name the isomers of butene.

16 Physical properties and trends of organic molecules

Summary

- The physical properties of organic molecules are a result of the intermolecular bonds that occur between them. The types of intermolecular forces that occur depend on the polarity of the bonds in the molecules and, therefore, the functional groups. The types of forces in order of increasing strength are summarised below.
  - Dispersion forces occur between all molecules and are the strongest type of interaction between non-polar molecules such as alkanes, alkenes and alkynes.
  - Dipole–dipole interactions occur between molecules with a polar bond, and are the strongest type of interaction between haloalkanes.
  - Hydrogen bonding exists between molecules with an O–H or N–H bond and is the strongest type of interaction between alcohols, carboxylic acids, amines and amides.
- Trends in melting and boiling points

  The melting and boiling points increase with the size of the molecules throughout every homologous series. As the number of carbon atoms increases, the surface area overlap between neighbouring molecules also increases; therefore, the strength of the dispersion forces acting is similarly greater. The molecules require more energy to be separated, and the melting and boiling points must be higher.

  This effect is also noticeable when comparing the melting and boiling points of isomers. Straight chain isomers have a greater surface area overlap and, therefore, a higher melting and boiling point than branched chain isomers with the same molecular mass.
- Trends in functional group

  Molecules with polar functional groups will have higher melting and boiling points than those without. Molecules such as alcohols that display hydrogen bonding will have higher melting and boiling points than haloalkanes; for example, which display only dipole–dipole interactions.
- Trends in solubility

  Solubility of organic molecules is a function of the extent to which the molecules can form positive interactions with water molecules. Water is a polar molecule that forms hydrogen bonds. Therefore, the greater the amount of polar bonds in an organic molecule, the greater the solubility will be. As such, more highly polar molecules such as carboxylic acids will dissolve in water to a greater extent than less polar molecules of the same carbon chain length.

  Non-polar carbon chains are described as hydrophobic as they do not form significant intermolecular bonds with water molecules. Polar functional groups such as hydroxyl or carboxyl groups do form positive hydrogen bonding interactions with water molecules and are described as hydrophilic. Therefore, the solubility decreases as the carbon chain length increases throughout homologous series such as alcohols and carboxylic acids, as the proportion of the molecule that is hydrophobic increases.

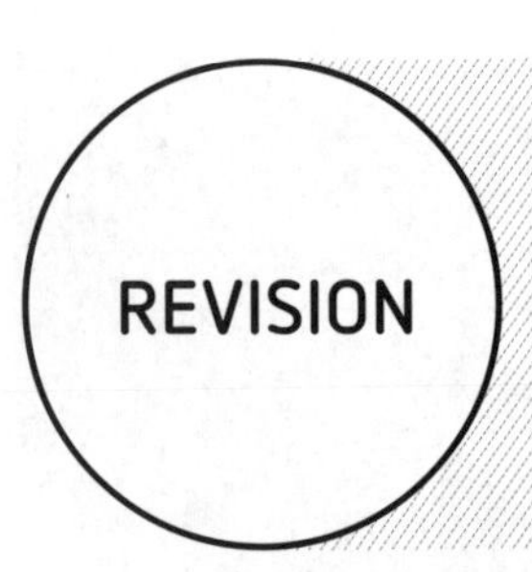

16.1 Important terms

1 Complete the crossword using the clues below.

Across	Down
3 The tendency of an atom to attract electrons from outside a covalent bond	**1** A molecule that does not bond attractively with water
6 A molecule that bonds attractively with water	**2** ______________ forces act between molecules that largely determine the physical properties of molecules.
7 ______________ forces: the weakest intermolecular forces that exist between all molecules	**4** _______ interactions: intermolecular forces that exist between molecules that have a permanent dipole
	6 ___________ bonds: strong intermolecular forces that exist between molecules that possess O–H, N–H or F–H bond
	7 A bond is said to be ___________ when there is a significant difference in electronegativity between the atoms of a covalent bond, and therefore an unequal distribution of electron density across the bond.

 9780170412476

2 Write a short summary or create a mind map using all of the terms from the crossword.

16.2 Compare and contrast

1 Which substance in the following pairs is more soluble in water and other polar solvents? Explain.

a Propanoic acid and hexanoic acid

b Pentane and 1,2-pentanediol

c Methyl methanoate and propyl propanoate

2 Would you expect that glycerol (1, 2, 3-propanetriol) is soluble in hexane? Explain.

3 Which substance in the following pairs would you expect to have the higher boiling point? Explain.

a 1-propene and 1-heptene

9780170412476

b Butane and 2-pentanol

____

____

____

____

____

____

c 1-propanol and 1, 2, 3-propanetriol

____

____

____

____

____

____

16.3 Solubility of alcohols

Listed in the table below are the solubilities of some alcohols and carboxylic acids, measured in grams of solute per 100 mL of water. (Note: miscible means completely soluble in water.)

TABLE 16.3.1 Solubilities of some alcohols and carboxylic acids

NAME	FORMULA	SOLUBILITY (g per 100 ml)
Methanol	CH_3OH	Miscible
Ethanol	CH_3CH_2OH	Miscible
Propanol	$CH_3CH_2CH_2OH$	Miscible
Butanol	$CH_3(CH_2)_2CH_2OH$	0.11
Pentanol	$CH_3(CH_2)_3CH_2OH$	0.030
Hexanol	$CH_3(CH_2)_4CH_2OH$	0.0058
Heptanol	$CH_3(CH_2)_5CH_2OH$	0.0008
Methanoic acid	$HCOOH$	Miscible
Ethanoic acid	CH_3COOH	Miscible
Propanoic acid	CH_3CH_2COOH	Miscible
Butanoic acid	$CH_3(CH_2)_2COOH$	Miscible
Pentanoic acid	$CH_3(CH_2)_3COOH$	5
Hexanoic acid	$CH_3(CH_2)_4COOH$	1.1

1 What do you notice about the relationship between solubility and the length of carbon chain? Explain the pattern that you observe, with reference to intermolecular bonding.

2 Use the data in the table to describe the difference in solubility between alcohols and carboxylic acids. Explain this difference with reference to intermolecular bonding.

3 Suggest how the solubility of the following substances would compare to that of the compounds shown in the table. Explain your reasoning.

a a chloroalkane

9780170412476

b an alkane

c an amine

Isomers of organic compounds share several features in common but differ in other features.

1 Which is of the following correctly describes the characteristics of a hydrogen compound?

A Different structures and different molar masses

B Identical physical properties but different chemical properties

C Identical empirical formulas but different boiling points

D Identical chemical properties but different products of complete combustion

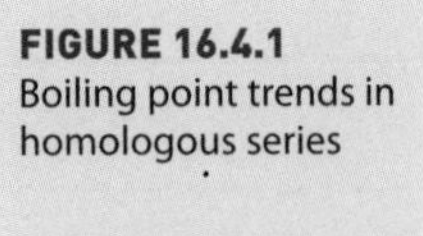

FIGURE 16.4.1
Boiling point trends in homologous series

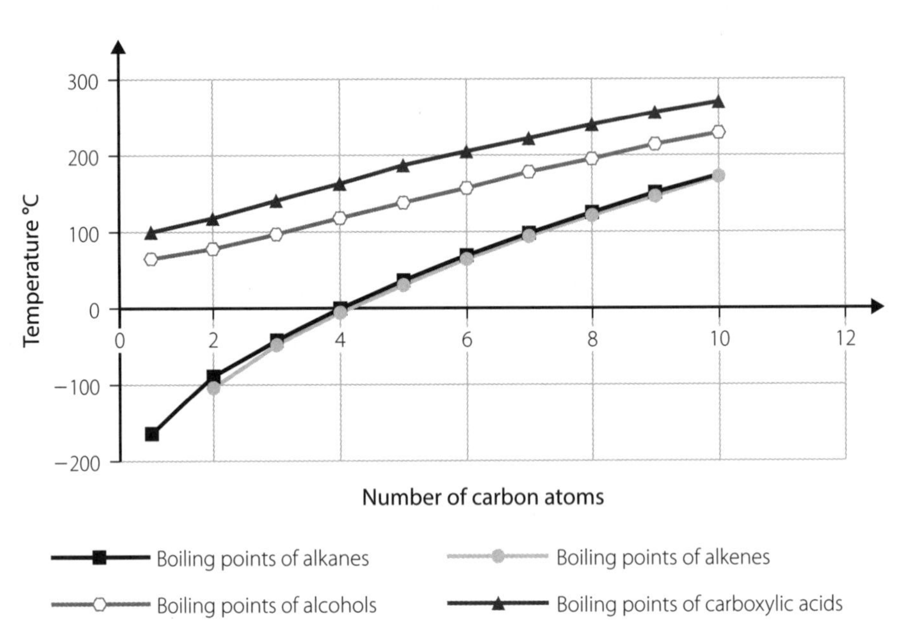

2 With reference to the bonding between molecules, answer the following questions.

a Why does the boiling point increase as the number of carbon atoms increases?

____

____

____

____

____

____

b Why are the boiling points of alcohols higher than the corresponding alkane?

c Why does the difference in boiling point between alcohol and alkene decrease as the number of carbon atoms in the chain increases?

17 Organic reactions and reaction pathways

Summary

Organic molecules participate in a variety of chemical reactions as a result of the functional groups they possess. Reactions take place under a specific set of conditions.

- Reactions of alkanes
  1. Combustion reactions

     Alkanes and all organic molecules can be burned in oxygen to produce carbon dioxide and water:

     $$CH_4(g) + 2O_2(g) \rightarrow CO_2(g) + 2H_2O(g)$$

     This reaction is known as complete combustion.

     If there is insufficient oxygen present, then incomplete combustion occurs, when carbon monoxide or carbon are formed together with water:

     $$2CH_4(g) + 3O_2(g) \rightarrow 2CO(g) + 4H_2O(g)$$

     or

     $$CH_4(g) + O_2(g) \rightarrow C(g) + 2H_2O(g)$$

  2. Free radical substitution reactions

     Alkanes and all organic molecules can react with halogen molecules in the presence of ultraviolet light. A substitution reaction occurs, where a hydrogen molecule is replaced (substituted) by a halogen atom on the original molecule:

     $$CH_4 + Cl{-}Cl \xrightarrow{\text{UV light}} CH_3Cl + H{-}Cl$$

     Methane | Chlorine | Chloromethane | Hydrogen chloride

     FIGURE 17.1.1 Free radical substitution reaction

- Reactions of alkenes

  Due to the presence of the C═C bond, alkenes take part in addition reactions, where two molecules combine together (add) to form a larger molecule. There are a number of reagents that can add to alkanes including hydrogen (H_2), halogen molecules (e.g. Cl_2), hydrogen halides (e.g. HBr) and water in the form of steam ($H_2O(g)$). Atoms from these reagents are added to each carbon of the double bond.

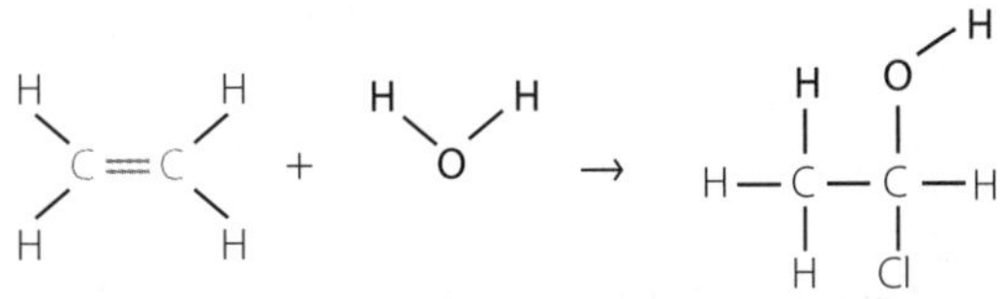

Ethene + Water → Ethanol

FIGURE 17.1.2 Addition reaction of ethene with water

If the alkene is unsymmetrical, then more than one product is possible.

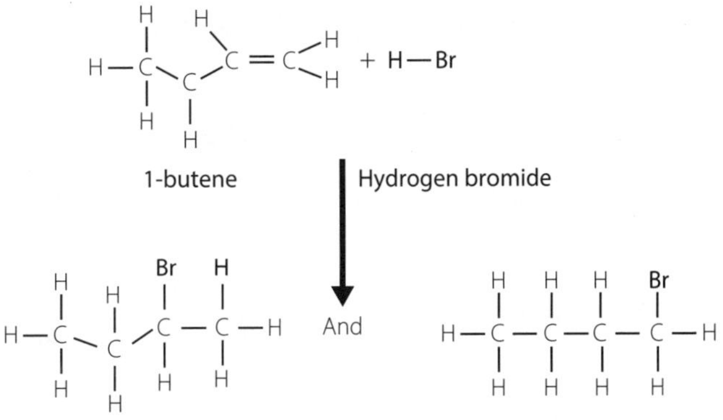

1-butene + H—Br (Hydrogen bromide) → 2-bromobutane And 1-bromobutane

FIGURE 17.1.3 Addition of HBr to propene

Markovnikov's rule states that in addition to hydrogen halides or water to unsymmetrical alkenes, the hydrogen atom is added preferentially to the alkene carbon atom, which is already bonded to the greater number of hydrogen atoms.

- Reactions of alkynes

  Alkynes take part in addition reactions in the same way as alkenes. As the alkyne contains a triple bond it reacts with twice as much reagent as alkenes in order to reach a saturated product.

$CH \equiv CH(g)$	+	$Br_2(aq)$	→	$CHBr = CHBr(g)$
Ethyne		Bromine		1,2-Dibromoethene
$CHBr = CHBr(g)$	+	$Br_2(aq)$	→	$CHBr_2CHBr_2(g)$
1,2-Dibromoethene		Bromine		1,1,2,2-Tetrabromoethane

FIGURE 17.1.4 Addition of bromine to alkynes

- Reactions of haloalkanes

  Haloalkanes take part in nucleophilic substitution reactions with electron-rich reagents known as nucleophiles. In these reactions, the halogen atom is replaced by the nucleophile.

Examples of nucleophiles include:

- aqueous hydroxide ions (e.g. NaOH) to form alcohols
- ammonia (NH_3) to form amines
- cyanide ions (e.g. KCN) to form nitriles.

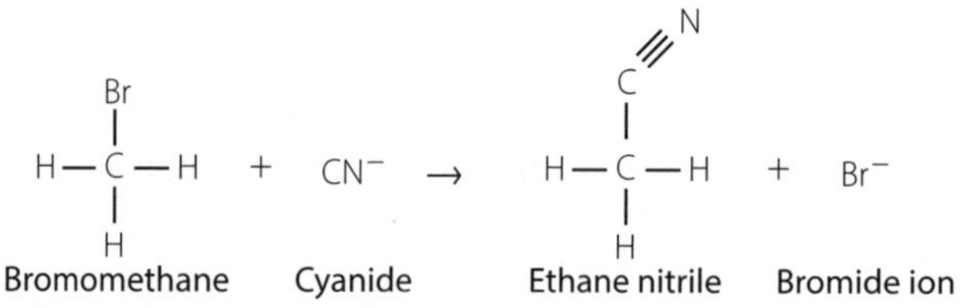

$CH_3Br + CN^- \rightarrow CH_3CN + Br^-$

Bromomethane, Cyanide, Ethane nitrile, Bromide ion

FIGURE 17.1.5 Reaction of bromomethane with potassium cyanide to form ethane nitrile

Haloalkanes also take part in elimination reactions, where a hydrogen halide molecule is lost (eliminated) to form an alkene molecule. The reagent for this is sodium hydroxide dissolved in ethanol.

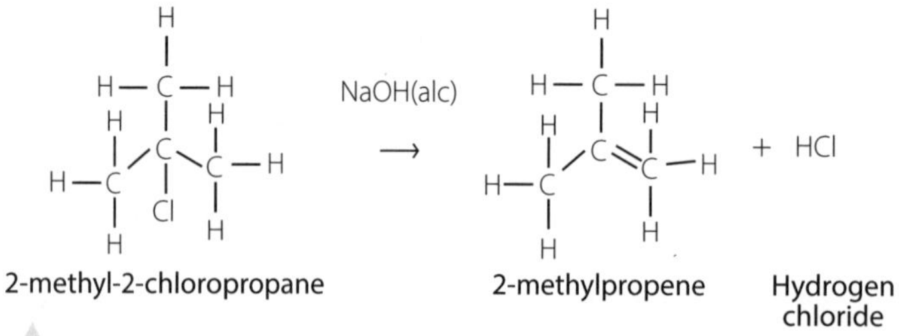

$(CH_3)_3CCl \xrightarrow{NaOH(alc)} (CH_3)_2C{=}CH_2 + HCl$

2-methyl-2-chloropropane, 2-methylpropene, Hydrogen chloride

FIGURE 17.1.6 Elimination reaction of 2-methyl-2-chloropropane to form the alkene 2-methylpropene

- Reactions of alcohols

Alcohols can be classified as primary, secondary or tertiary:

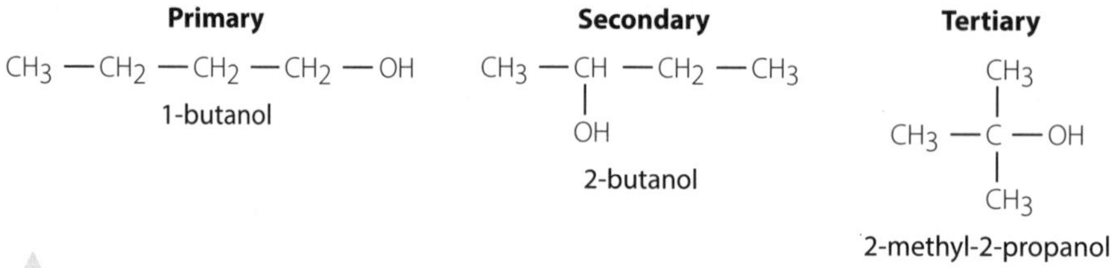

Primary	**Secondary**	**Tertiary**
$CH_3-CH_2-CH_2-CH_2-OH$	$CH_3-CH(OH)-CH_2-CH_3$	$(CH_3)_3C-OH$
1-butanol	2-butanol	2-methyl-2-propanol

FIGURE 17.1.7 Primary, secondary and tertiary alcohols

Alcohols can take part in combustion and elimination reactions as described previously. However, they can also take part in oxidation reactions, with a reagent such as acidified potassium permanganate(VII) ($KMnO_4$) or acidified potassium dichromate(VI) ($K_2Cr_2O_7$) depending on their structure:

Primary alcohols are oxidised to form aldehydes, which are in turn oxidised to form carboxylic acids.

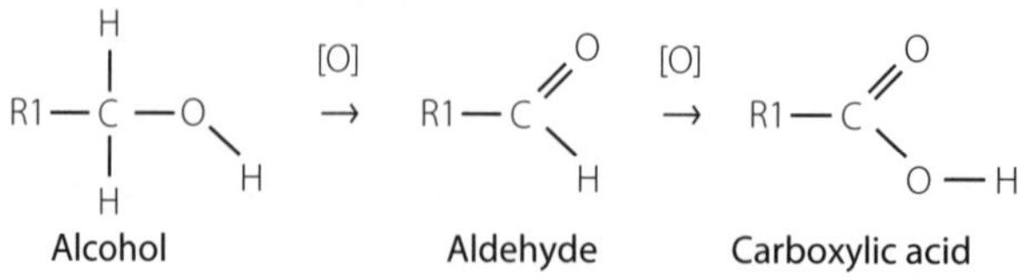

$R1-CH_2-OH \xrightarrow{[O]} R1-CHO \xrightarrow{[O]} R1-COOH$

Alcohol, Aldehyde, Carboxylic acid

FIGURE 17.1.8 Oxidation of primary alcohols (R1 indicates the alkyl section of the molecule). Secondary alcohols are oxidised to form ketones, which cannot be further oxidised.

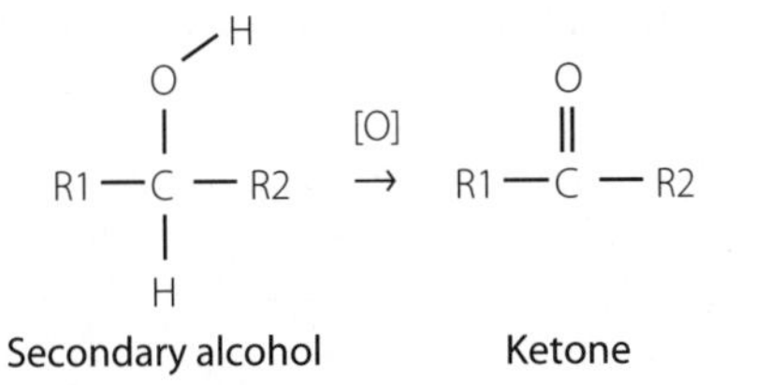

$$R1-CH(OH)-R2 \xrightarrow{[O]} R1-CO-R2$$

Secondary alcohol   Ketone

FIGURE 17.1.9 Oxidation of secondary alcohols

The overall and half-equations for the oxidation of 2-propanol with acidified potassium manganite(VII) are:

Oxidation: $5CH_3CHOHCH_3(aq) \rightarrow 5CH_3COCH_3\,(aq) + 10H^+(aq) + 10e^-$

Reduction: $2MnO_4^-(aq) + 16H^+(aq) + 10e^- \rightarrow 2Mn^{2+}(aq) + 8H_2O(I)$

Overall: $5CH_3CHOHCH_3(aq) + 2MnO_4^{2}(aq) + 6H^{1}(aq) \rightarrow 5CH_3COCH_3(aq) + 2Mn^{2+}(aq) + 8H_2O(I)$

Tertiary alcohols do not react with oxidising agents.

- Alcohols can also react with carboxylic acids in a condensation reaction, to form esters. This reaction (also known as esterification) is moderately slow at room temperature. To speed up the reaction, the mixture is heated in the presence of an acid catalyst, usually concentrated sulfuric acid. The production of water classifies this reaction as condensation.

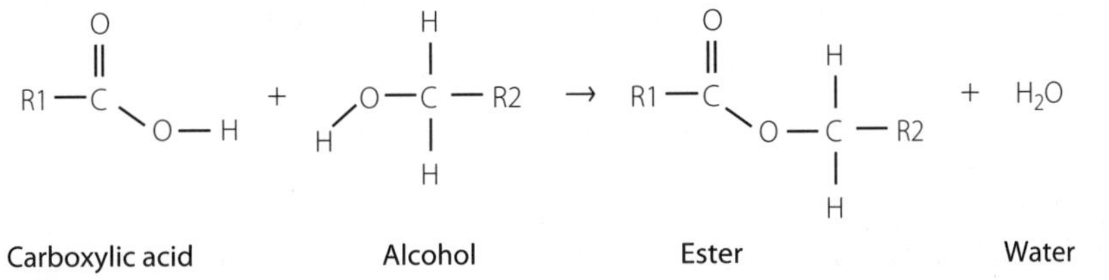

$$R1-COOH + H-O-CH_2-R2 \rightarrow R1-COO-CH_2-R2 + H_2O$$

Carboxylic acid   Alcohol   Ester   Water

FIGURE 17.1.10 The formation of an ester

- Esters can also be hydrolysed to reform the carboxylic acid and the alcohol in a reaction catalysed by either acid or base. Under acid conditions (in the presence of heat and sulfuric acid), the reaction is as shown in Figure 17.1.11.

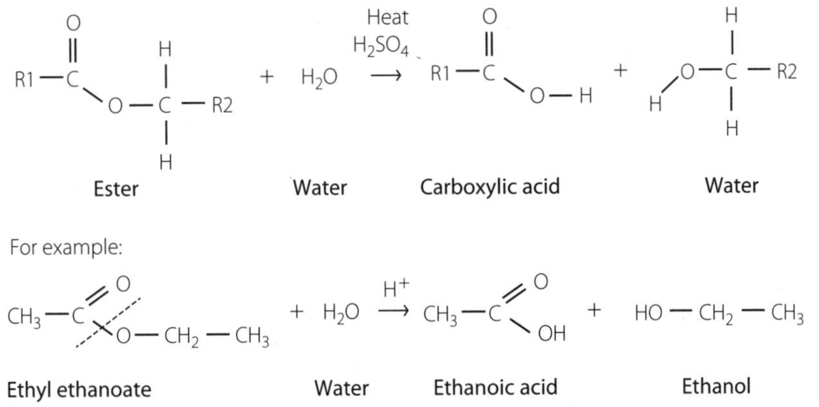

$$R1-COO-CH_2-R2 + H_2O \xrightarrow[H_2SO_4]{Heat} R1-COOH + H-O-CH_2-R2$$

Ester   Water   Carboxylic acid   Water

For example:

$$CH_3-COO-CH_2-CH_3 + H_2O \xrightarrow{H^+} CH_3-COOH + HO-CH_2-CH_3$$

Ethyl ethanoate   Water   Ethanoic acid   Ethanol

FIGURE 17.1.11 The acid hydrolysis of an ester

Under alkaline conditions (in the presence of heat and sodium hydroxide), the reaction is as shown in Figure 17.1.12.

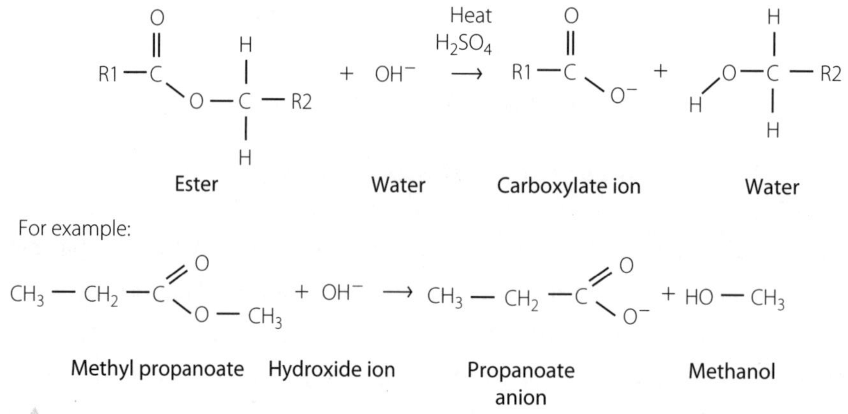

$$R1{-}C({=}O){-}O{-}CH_2{-}R2 + OH^- \xrightarrow[H_2SO_4]{Heat} R1{-}C({=}O){-}O^- + HO{-}CH_2{-}R2$$

Ester Water Carboxylate ion Water

For example:

$$CH_3{-}CH_2{-}C({=}O){-}O{-}CH_3 + OH^- \longrightarrow CH_3{-}CH_2{-}C({=}O){-}O^- + HO{-}CH_3$$

Methyl propanoate Hydroxide ion Propanoate anion Methanol

FIGURE 17.1.12 The alkaline hydrolysis of an ester

- Carboxylic acids are weak acids compared to inorganic acids such as HCl and H_2SO_4. However, they still react in the same way as these acids. Typical reactions include:
  - Reaction with reactive metals

    Carboxylic acids react with reactive metals to produce a salt and hydrogen gas.

    For example:

    $$Mg(s) + 2CH_3COOH(aq) \rightarrow Mg^{2+}(aq) + 2CH_3COO^-(aq) + H_2(g)$$
  - Reaction with base (neutralisation)

    Carboxylic acids react with a base to produce a salt and water in a neutralisation reaction.

    For example:

    $$CH_3COOH(aq) + NaOH(aq) \rightarrow Na^+(aq) + CH_3COO^-(aq) + H_2O(l)$$
- Reactions of amines.

  Amines are weak bases, and so react with inorganic acids.

  For example:

  $$CH_3{-}NH_2(aq) + HCl(aq) \rightarrow CH_3 - NH_3 + Cl^-(aq)$$

  Methanamine Methylammonium chloride

  Amines can also react with carboxylic acids in a condensation reaction in order to form amides.

  For example:

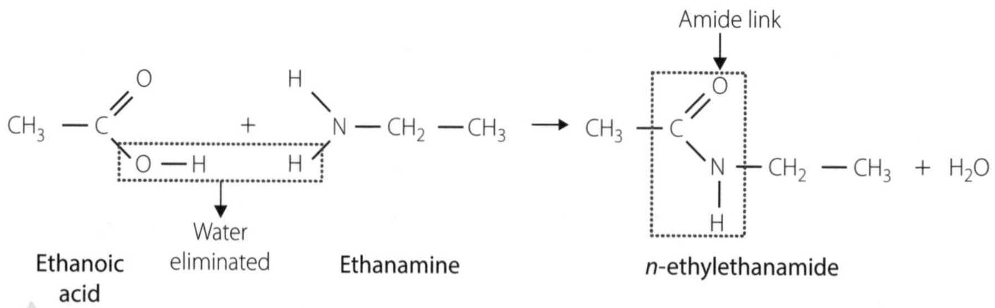

$$CH_3{-}C({=}O){-}OH + H{-}N(H){-}CH_2{-}CH_3 \longrightarrow CH_3{-}C({=}O){-}N(H){-}CH_2{-}CH_3 + H_2O$$

Amide link

Water eliminated

Ethanoic acid Ethanamine *n*-ethylethanamide

FIGURE 17.1.13 The condensation reaction of an amide

17.1 Important terms

1 Complete the crossword using the clues on page 145.

using the clues on page 145

Across	Down
6 An elimination reaction where a molecule of water is lost from an alcohol to form an alkene	**1** A hydrogenation reaction to either alkenes or nitriles
9 Free ________ substitution: a UV light-initiated reaction where a H atom is replaced with a halogen	**2** A reaction where a larger molecule is split into two smaller molecules by reacting with water and an acid or base catalyst
13 ____________ combustion: burning of an organic compound in limited oxygen to produce carbon monoxide or carbon and water	**3** An acid that dissociates producing one hydrogen ion per molecule
16 The —COOC— group that links two alkyl chains in an ________ molecule	**4** Unsaturated molecules combining with another molecule to become saturated
18 ________ alcohol is an alcohol where the —COH group is joined to one other carbon atom.	**5** __________ alcohol is where the —COH group is joined to two other carbon atoms.
19 A reaction where an oxidising agent is used to convert primary alcohols to aldehydes and carboxylic acids, and secondary alcohols to ketones	**7** Reaction where a molecule is lost from a larger molecule, producing an unsaturated product
20 __________ distillation: separation of a mixture of organic compounds according to their different boiling points	**8** Addition reaction involving a halogen molecule
	10 ____________ reaction: a reaction where two molecules combine, forming a larger one and produce water as a by-product
	11 ________ alcohol: an alcohol where the —COH group is joined to three other carbon atoms
	12 __________ combustion: burning of an organic compound in sufficient oxygen to produce carbon dioxide and water
	14 ___________ substitution: reaction of haloalkanes where the halogen atom is replaced by an electron rich species called a nucleophile
	15 The condensation reaction where an ester is formed from an alcohol and a carboxylic acid
	17 An addition reaction where hydrogen is the molecule added

2 Write a short summary or create a mind map using all of the terms from the crossword.

9780170412476

17.2 Reaction types

Complete the table below to identify and highlight the key features of each type of organic reaction.

	DESCRIPTION	EXAMPLE, WITH REAGENTS AND CONDITIONS
Addition		
Combustion		
Condensation		
Elimination		
Oxidation		
Reduction		
Substitution (free radical)		
Substitution (nucleophilic)		

17.3 Drawing and identifying structures and reactions

1 Draw the structures of the molecules of:

a a primary alcohol

b a secondary alcohol

c a tertiary alcohol.

2 Write the systematic names of the alcohols you drew in question 1.

3 Compare the reactivity and oxidation products from the three alcohols that you drew in question 1, when acidified dichromate solution is added to samples of each.

4 Write an equation for the hydrolysis, under acidic conditions, of propyl ethanoate. Name the products of this reaction.

5 For each of the reactions represented below, state whether it is an addition reaction, a substitution reaction or a condensation reaction. Write an equation for each reaction.

a Reaction of propanoic acid and methanol to form an ester

b Reaction of 1-propene with chlorine in acidic conditions to form 1, 2-dichloropropane

c Reaction of butane with bromine, with ultraviolet radiation, to form 1-bromobutane

17.4 Identifying reactions

1 For each of the molecules below, identify a reaction other than combustion that it will take part in and write an equation to represent that reaction.

a

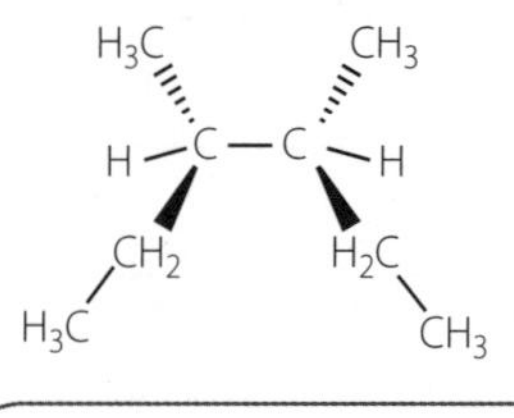

b

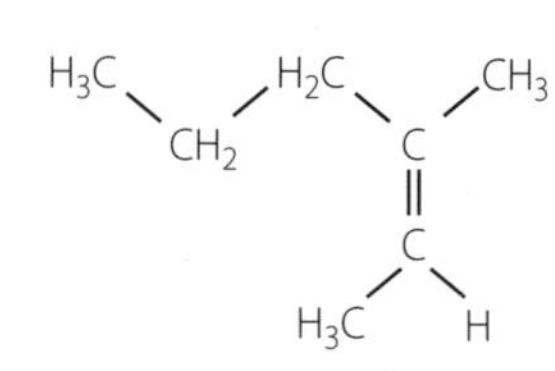

c

H3C H2C CH3
CH2 CH
OH

d

HO H2C — CH3
H C H
H3C H

e

H3C H2C H2C
CH2 CH2 COOH

9780170412476

Butane and propane can be used to produce the ester 2-propyl butanoate. The reaction scheme below shows the stages in this process.

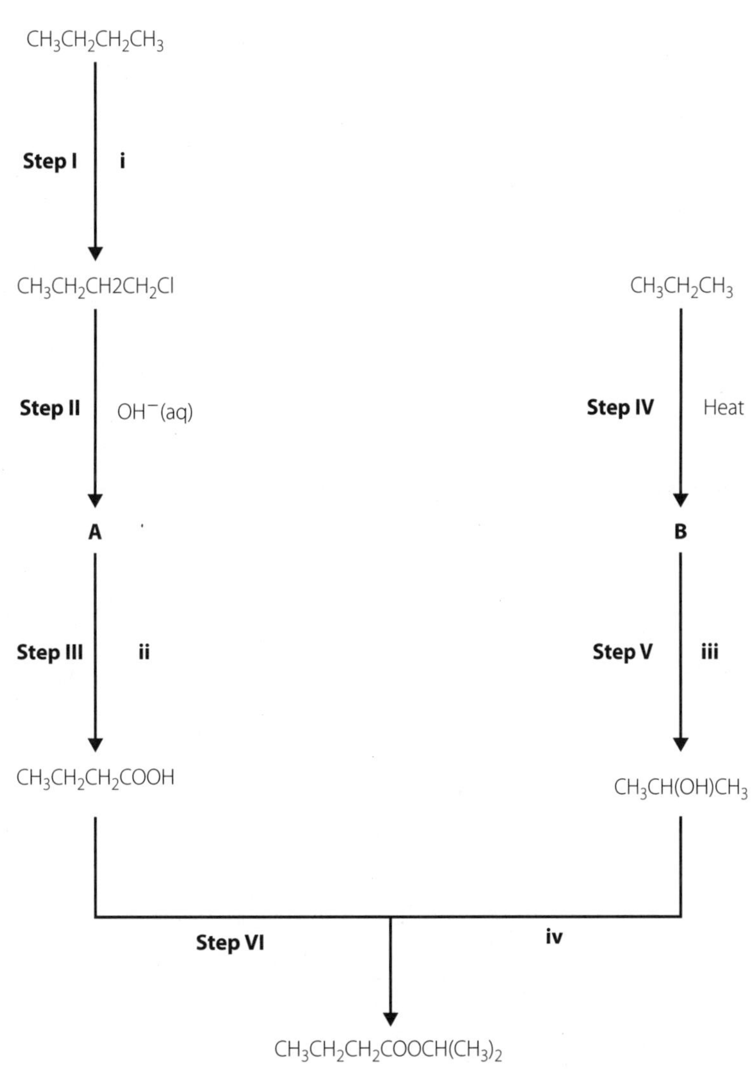

1 What are the semi-structural formulas for the compounds labelled **A** and **B**?

A: ______________________________

B: ______________________________

2 What are the additional reactants **i**, **ii**, **iii** and **iv**?

i:

ii:

iii:

iv:

3 What types of chemical reaction occur in the following steps?

Step II:

Step III:

Step IV:

Step V:

Step VI:

4 What are the systematic names for the compounds produced in the following steps?

Step III ($CH_3CH_2CH_2COOH$)

Step IV ($CH_3CH(OH)CH_3$)

5 What are the by-products of the reaction in **Steps II**, **IV** and **VI**?

Step II:

Step IV:

Step VI:

18 Organic materials: Structure and function

LEARNING

Summary

- Many organic molecules with biological significance, such as proteins, carbohydrates and lipids, are polymers, macromolecules formed by the linking of many thousands of smaller molecules known as monomers. The properties of these molecules and other synthetic polymers can be explained by understanding how these chains form.
- Polymers, both naturally occurring and synthetic, vary in terms of their properties such as biodegradability, tensile strength and density.
- Proteins are polypeptide molecules formed from linking α-amino acid monomers. The monomers are held together through the formation of peptide (amide) bonds between the amino group from one monomer and the carboxyl group of the next.
- Protein structure can be described in a series of stages:
  - Primary structure – the sequence of amino acid monomers in the chain
  - Secondary structure – the formation of structures such as α-helices and β-sheets in a peptide chain held together by intermolecular hydrogen bonding between side chains in the amino acid molecules
  - Tertiary structure – the three-dimensional structure formed by a single peptide as a result of the accumulated intermolecular interactions between functional groups in the amino acid monomers. This is very important for the biological function of the protein.
  - Quaternary structure – the aggregation of two or more tertiary structures to form a unit, held together by intermolecular interactions. Again, the quaternary structure of a protein is critically important in determining how a protein interacts with other molecules and, therefore, its biological reactivity.
- Enzymes are protein molecules that act as catalysts for biological reactions. They lower the activation energy of reactions by providing a lower energy pathway by which the reaction can occur. This can be done by binding to substrate molecules through the enzyme's active site and holding them in a particular conformation that enables the reaction to take place more easily.
- Enzymes, like all proteins, will undergo a permanent change to their structure: denaturing. If the external conditions such as temperature and pH are changed to the extent that they cause the intermolecular interactions within the protein structure to change. As such, the activity of enzymes is significantly affected by changes in temperature and pH.

- Carbohydrates are a large group of organic molecules with the general formula CH_2O.
    - The smallest carbohydrate molecules are the monosaccharides. These include the aldose sugars such as glucose and galactose, which contain an aldehyde group, and the ketose sugars such as fructose, which contain a ketone group.

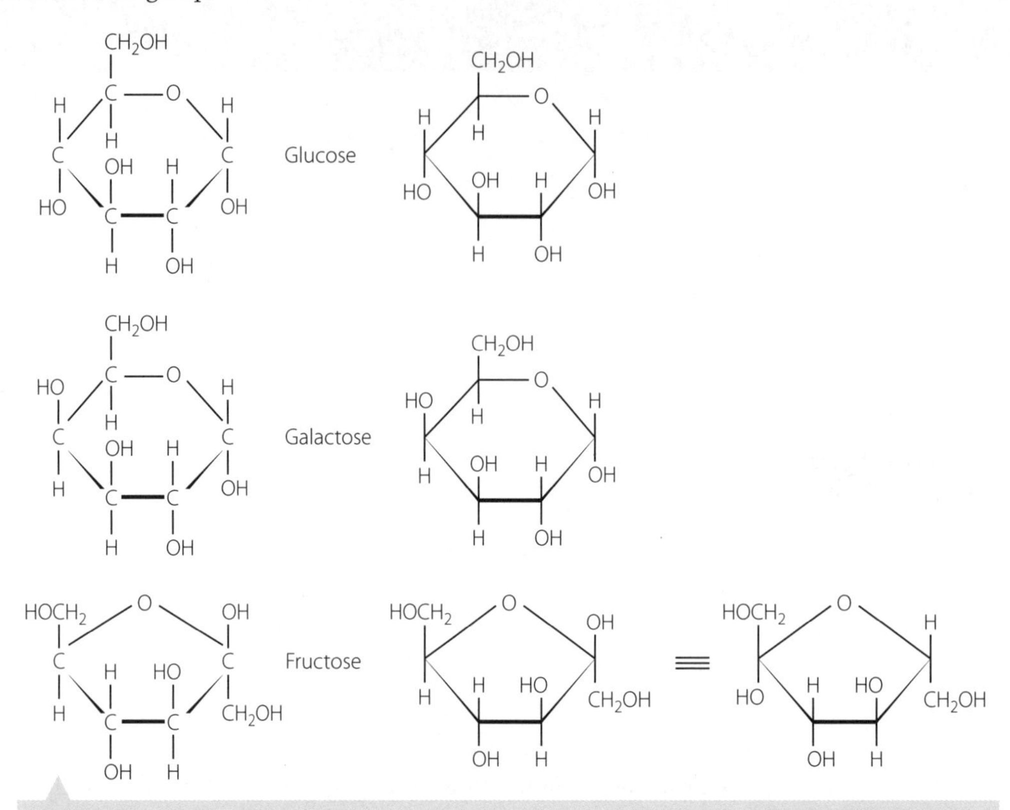

FIGURE 18.1.1 Structures of three common monosaccharides. The left-hand structures show all the atoms of the molecules; in the right-hand ones, the carbon atoms are not labelled. The right-hand structure for fructose is the middle structure flipped through 180°.

- The aldose and ketose sugars are all isomers of one another, with the general formula $C_6H_{12}O_6$. Their structures are therefore very similar. Glucose also has two biologically active forms, α-glucose and β-glucose, the only difference between the two is the orientation of —OH functional groups. However, this difference in structure results in significantly different biological activity.

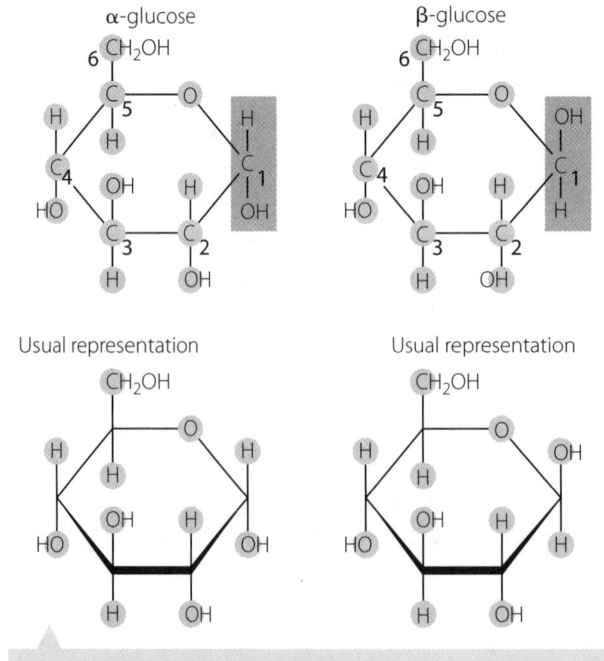

FIGURE 18.1.2 The α- and β- forms of glucose

- Starch and cellulose are examples of polysaccharides, long-chain polymers of monosaccharide molecules, held together by ether (—C—O—C—) bonds known as glycosidic links.
- Starch is formed from α-glucose, and can take two forms: amylose, where the polymer chains are straight, and amylopectin, where the polymer chains are highly branched.
- Cellulose is formed from β-glucose, and as a result cannot be digested by animals. It is an unbranched structure but with significant cross-linking from intermolecular hydrogen bonds. It is a highly rigid, crystalline substance that is used by plants to form stems and leaves, where strength and rigidity are essential.

- Lipids are triglyceride molecules, esters formed from the reaction of fatty acid molecules with glycerol (propane-1,2,3-triol).
- Fatty acid molecules consist of a carboxyl group at the end of a long hydrocarbon chain. Unsaturated fatty acid molecules have no C=C bonds within the chain, whereas unsaturated fatty acid molecules have one or more C=C bonds.
- Lipid molecules can be hydrolysed in a reaction catalysed by an alkali, such as sodium hydroxide, to produce glycerol and the sodium salt of the fatty acid.

$CH_3(CH_2)_{16}COO^-Na^+$

FIGURE 18.1.3 Structure of a soap molecule, showing the long hydrocarbon tail and carboxylate ion head

These molecules can be used as soaps, as the long, non-polar carbon chain readily mixes with insoluble molecules, and the charged carboxylate ion ensures that the molecule dissolves in water.

- The structure of synthetic polymers is also critical to understanding their properties. Examples of these include high-density polyethene (HDPE) and low-density polyethene (LDPE).
  - HDPE consists of straight polymer chains with no branching. As a result, the molecules form a crystalline structure, where the chains are closely packed. HDPE is a harder, more rigid substance with a higher melting point than LDPE.
  - LDPE consists of polymer chains with significant branching. As a result, the chains do not pack as closely together. Hence, LDPE is a soft, flexible molecule with a low melting point.
- Another example of variation of structures in synthetic polymers is that of polypropene. In this case, the relative position of the methyl groups in the polymer chain results in three polymers, each with different properties: syntactic, isotactic and atactic polypropene.

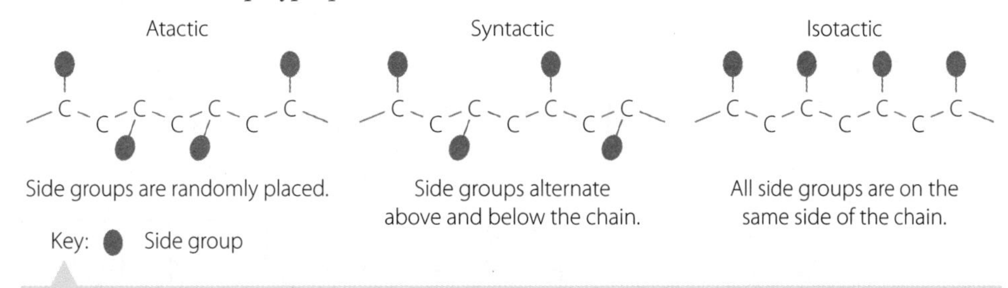

FIGURE 18.1.4 Three possible arrangements of side groups in polypropene

- Polytetrafluoroethene (PTFE) has highly polar carbon-fluorine bonds. As such, it forms strong dipole-dipole forces between molecules and is resistant to the formation of dispersion forces. Therefore, it has a very low coefficient of friction, which makes it very useful as a lubricant and for forming non-stick surfaces.

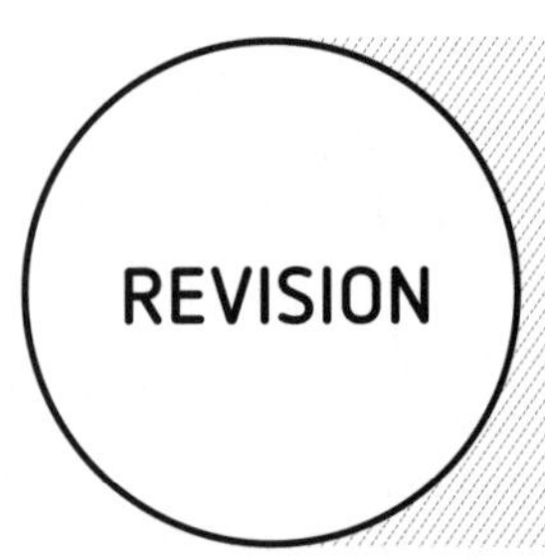

18.1 Important terms

1 Complete the crossword using the clues on page 156.

Across	Down
3 A polymer made from more than one monomer **4** _____ acid: an organic compound containing a carboxyl and an amino group **5** The mass per unit volume **8** An organic compound made from long chains of amino acids **9** A large molecule made from many thousands of repeating units **12** ________ strength: the resistance of a substance to being pulled or stretched **15** The form of an amino acid where the amino group is protonated at the same time as the carboxyl group is deprotonated, so that there is no net charge **16** _____ groups: the different functional groups that can exist in a polymer chain, but which are not part of the main chain **17** A substance that will decay naturally as a result of enzyme produced by living bacteria	**1** A polymer held together by amide bonds, formed between the amino group of one monomer and a carboxyl group of the next monomer **2** The process by which monomers link together to form a polymer **4** Covalent bonds that exist between polymer chains, holding them together firmly **6** Where a polymer chain is not a single straight chain, but with 'offshoots' **7** A reaction where a water molecule is formed as a by-product **8** A polyamide chain that constitutes a protein **10** __________ structure: the shape of two or more polymer chains aggregated together, held by intermolecular interactions **11** A reaction where alkenes lose their double bonds to form a saturated molecule **13** ___________ structure: minor features of the shape of a polymer chain, caused by intermolecular interactions between side chains **14** An individual unit of a polymer **18** ___________ point: pH value at which the amino acid is in the zwitterion form

2 Write a short summary or create a mind map using all of the terms from the crossword.

This page has been left blank to complete the dominoes activity.

9780170412476

18.2 Dominoes

Cut out each of the dominoes below. Starting with the 'Beginning' domino, place the correct answer to the right of each question or statement until you finish with the 'End' domino.

Beginning	All living things are massive chemical machines composed largely of
have a double bond between adjacent carbon atoms.	What are the monomer units in polyethylene?
silk, wool and hair.	The characteristic of thermoplastic polymers is that
form polyamides	in which the monomer units are joined by
the sequence of amino acids along the protein chain.	Hydrogen bonding between amino acids on the same side of different chains
are called plasticisers.	Polymers that have more than one monomer in the chains
are molecules of glucose, a monosaccharide.	Sucrose, maltose and lactose are
They are all on the same side of the polymer chain.	Because of the random distribution of side chains in atactic polymers,
amino acids.	Both the carboxylate and the amine groups on amino acids are protonated if
they soten on heating and can be moulded.	Polymers that do not soften as the temperature is raised
they are in low pH conditions.	The primary structure of a protein is
the amide link, —CO—NH—.	The monomer units in the carbohydrates cellulose and starch
attraction between chains is low and the polymers are soft.	Non-polymer molecules that reduce attractions between chains
gives rise to two- or three-dimensional structures that are more rigid.	Condensation polymers are formed from
without loss of atoms from the monomers.	The monomers from which addition polymers are formed

This page has been left blank to complete the dominoes activity.

9780170412476

organic compounds interacting in a watery soup.	Polymers are composed of giant molecules formed from	are called copolymers.	Cross-linking that joins chains together	they can be continually produced from crops or algae.	The production of ethanol from sugars, in the presence of yeasts, is called
disaccharides.	Because disaccharide molecules are too big to pass through cell membranes	polyesters are formed.	Monomers with both a carboxylic acid and an amine	gives rise to the secondary structure of proteins.	The tertiary structure of a protein is
because chain branching prevents close packing of the molecules.	Because of the absence of branching in HDPE,	are called thermosetting polymers.	Addition polymerisation occurs	small molecules (monomers) joined by covalent bonds.	Natural polymers that are proteins include
the overall shape, resulting in the folding of the helices or sheets of the secondary structure.	Biofuels are regarded as renewable energy sources because	the packing of molecules is regular, and called crystalline.	What is the arrangement of side chains in isotactic linear polymers?	they must be broken down to the monosaccharides for digestion.	The monomer units of proteins are
monomers that have two different functional groups.	From monomers with both a carboxylic acid and an alcohol,	Ethene (ethylene) molecules, $CH_2{=}CH_2$	Low-density polyethylene has weak intermolecular forces	fermentation.	**End**

This page has been left blank to complete the dominoes activity.

18.3 Explaining the properties of fats and oils

Melting point is one of the properties of substances that depend on the strengths of forces of attraction between molecules.

Fats and oils are composed of triglyceride molecules: esters of glycerol with carboxylic acids that have long-chain hydrocarbon 'tails'. There are about 40 different carboxylic acids that occur naturally in triglycerides, with different concentrations of each in triglycerides from different sources. The melting points of triglycerides depend on the intermolecular forces between these hydrocarbon tails on different molecules. The visual difference between fats (solids) and oils (liquids) is because oils are above the melting points of the triglycerides.

1 Which have the higher melting points: the triglycerides in fats, or those in oils?

The main difference between the hydrocarbon tails is that some are saturated chains, and some are unsaturated, with one or more double bonds. An example of a carboxylic acid with an unsaturated hydrocarbon chain is stearic acid (Figure 18.3.1), and one with three double bonds in the chain is linolenic acid (Figure 18.3.2).

$$CH_3CH_2CH_2CH_2CH_2CH_2CH_2CH_2CH_2CH_2CH_2CH_2CH_2CH_2CH_2CH_2CH_2C(=O)OH$$

FIGURE 18.3.1 Stearic acid

$$CH_3CH_2CH{=}CHCH_2CH{=}CHCH_2CH{=}CHCH_2CH_2CH_2CH_2CH_2CH_2CH_2C(=O)OH$$

FIGURE 18.3.2 Linolenic acid, a polyunsaturated fatty acid (PUFA)

Using molecular models, or materials used to make molecular models, construct a few models each of stearic acid and linolenic acid. Pack the stearic acid models side by side, and do the same with models of linolenic acid molecules.

2 What is the difference in the ability of stearic acids and linolenic acids to pack closely together?

3 Which molecules pack most closely?

4 Which might you expect to have the stronger intermolecular attractions between the adjacent molecules?

5 Imagine a triglyceride in which all of the carboxylic acids are stearic acid, and another in which all are linolenic acid, and that one is a fat and the other an oil. Which one would you expect to be the fat, and which the oil?

6 Explain why a common process on an industrial scale in the food industry is to hydrogenate the unsaturated hydrocarbon tails of some triglycerides.

9780170412476

18.4 Comparing and contrasting biological polymers

1 Complete the table to compare and contrast essential information about proteins, lipids and carbohydrates.

NAME	MONOMER UNIT(S)	NAME OF BOND LINKING MONOMERS	BIOLOGICAL ACTIVITY	DESCRIPTION OF STRUCTURE
Proteins				
Saturated lipids				
Unsaturated lipids				
Cellulose				
Starch (amylose)				
Starch (amylopectin)				

EVALUATION

1 a Egg white contains the large protein, albumin, which has a characteristic three-dimensional tertiary structure. Define 'tertiary structure' and explain how this three-dimensional structure depends upon the sequence of amino acids in the primary structure.

b Describe one way in which the denaturation of the protein albumin can occur.

c Explain why the structure of proteins in enzymes enables them to be far more efficient catalysts than those that can be developed in a laboratory.

9780170412476

2 Cellulose is a polysaccharide sugar, made from glucose molecules. Figure 18.5.1 shows the structure of cellulose.

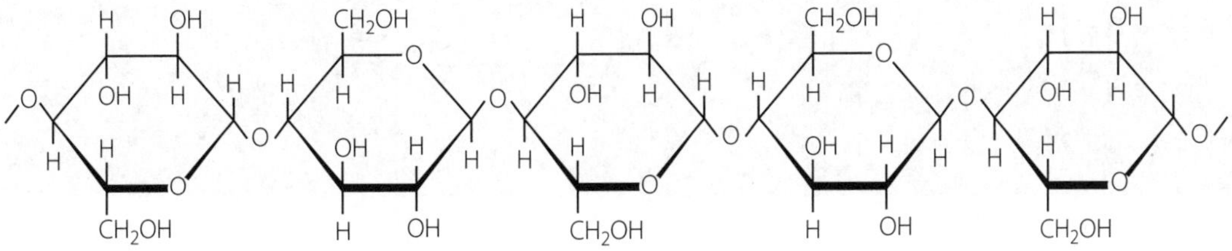

FIGURE 18.5.1 Cellulose

a Draw a diagram to represent a glucose molecule.

b Name the type of bond that joins glucose molecules together in cellulose.

____

c Explain why cellulose cannot be digested by animals; whereas starch, another polymer of glucose, can be digested by animals.

____

____

____

3 Vegetable oils are triglyceride esters, which can be broken down to form polyunsaturated fatty acids. This is represented by the equation shown.

$$\begin{array}{l} CH_2\text{–}O\text{–}CO\text{–}C_{17}H_{31} \\ | \\ CH\text{–}O\text{–}CO\text{–}C_{17}H_{31} \quad + \ 3\mathbf{A} \rightarrow 3C_{17}H_{31}COOH \ + \ \mathbf{B} \\ | \\ CH_2\text{–}O\text{–}CO\text{–}C_{17}H_{31} \end{array}$$

a Suggest identities for chemicals **A** and **B**.

A: ______________________________

B: ______________________________

b Explain why the fatty acid product shown could be described as a polyunsaturated fatty acid.

____

____

19 Analytical techniques

Summary

- In order to determine the structure of complex organic molecules such as proteins, a number of different instrumental techniques can be used. In practice these techniques are often used in conjunction to obtain the necessary information.
- *Chromatography* is a technique used to separate and quantify the different components in a mixture. It is commonly set up in the form of a column. The sample is placed on top of the column, filled with a substance known as the stationary phase. The mobile phase is then passed down the column and carries the sample with it. The different components in the sample then separate according to their varying affinity with the stationary and mobile phases. As they are separated, each component will vary in retention time, that is, the time taken to pass through the column.

  There are several variations of this technique used for protein analysis in particular, including:
  - Ion exchange chromatography – the pH is adjusted so that the protein molecules are charged and then absorb strongly to either a cation or anion exchange resin as the stationary phase.
  - Size exclusion chromatography – the protein molecules can pass through pores in the stationary phase, depending on their size. In this case, retention time is inversely proportional to the size of the molecules.
  - Hydrophobic interaction chromatography – the stationary phase has a very high salt concentration. This makes the sample less soluble and therefore causes adsorption according to the non-polar, hydrophobic regions of the protein.
- *Electrophoresis* is a technique similar to that of chromatography, where charged particles are separated on the basis of their varying rates of migration through a medium such as a gel. However, there is no mobile phase; the migration is the result of an electric field being applied across the gel, causing a force to act on the charged particles. Because protein molecules are charged as a result of the polar amino and carboxyl functional groups in the peptide bonds and in the side chains, electrophoresis is a particularly useful technique for protein analysis.
- *Mass spectrometry* involves the ionisation and fragmentation of larger molecules, and the separation and analysis of these fragments. It is used to provide information about the structure of the original molecule. The mass spectrum produced is a graphical representation of the fragmented mixture, representing the different fragments as peaks on the graph, each with a different mass: charge ratio. The relative height of each peak corresponds to the relative abundance of each fragment.
  - When fragmentation occurs, species called radicals and ions are formed. Fragmentation involves the breaking of a covalent bond, which is a shared pair of electrons. One of these shared electrons is lost, as the positively charged ion is formed. The other electron remains in an

9780170412476

excited state on the other half of the molecule, forming what is known as a radical. Radicals are uncharged and therefore cannot be detected by the mass spectrometer. Often, the fragment with the more electronegative elements will form the radical preferentially and the less electronegative fragment will form the ion.
- Therefore, analysis of the mass spectrum reveals the formation of characteristic fragments. Depending on bond strengths, some bonds are more likely to fragment than others. Conclusions can then be drawn about the structure of the original molecule. Mass spectrometry is most often used in conjunction with other techniques to determine the structure of an organic molecule.

- *X-ray crystallography* is another technique that can be used to assist in the determination of a molecular structure, and in particular, for proteins, enzymes and viruses.
  - First a crystal of the molecule being analysed is obtained. A crystal is a solid containing a regular array of the molecules arranged in a repeating pattern.
  - Next X-rays are shone through the crystal. Because the wavelength of the X-ray is similar to the bond lengths between the atoms, the X-rays are scattered and a diffraction pattern is obtained.
  - Analysis of the diffraction pattern provides evidence about the distances between atoms in the molecule and also the three-dimensional geometric arrangement of the atoms. As a result, the bond lengths and bond angles can be determined.
- *Infrared (IR) spectroscopy* measures the absorption by the molecule of specific frequencies of IR radiation. These frequencies are measured in terms of wavenumbers in cm^{-1}, the number of waves that can occur per cm. IR spectroscopy is also known as vibrational spectroscopy, because absorption of these frequencies causes individual bonds to vibrate. Different bonds (e.g. C—C, C—H, C—O, O—H or C=O) each have their own characteristic frequencies of absorption and so the presence of these bonds can be determined by analysis of the IR absorption spectrum for a molecule. The absorption of specific bonds can be clearly determined in the region of 1500 – 3500 cm^{-1}, where some very characteristic absorptions associated with specific bonds can be identified. The region from 800–1500 cm^{-1} is known as the 'fingerprint' region. It is not used for the identification of specific bonds, but for comparison with a library sample to identify the molecule as a whole.

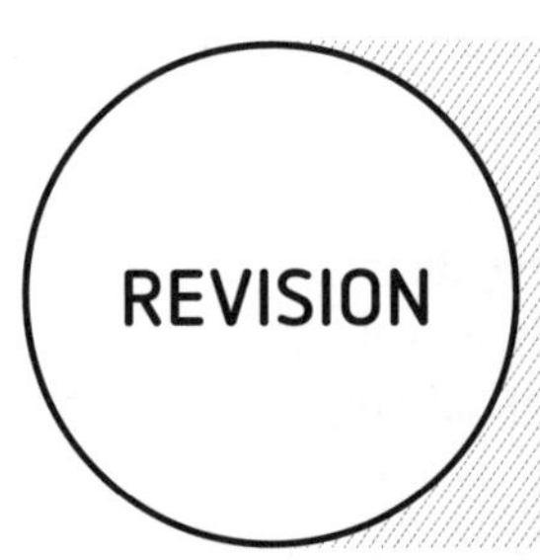

19.1 Important terms

1 Complete the crossword using the clues on page 169.

Across	Down
3 Constructive ____________: two waves mutually reinforcing each other, producing a combined wave with amplitude equal to the sum of the constituent waves **4** A species formed with an unpaired, high energy electron in an excited state **8** One of the substances being separated from a mixture by chromatography **9** Gel ______________: the process of separating large charged molecules by placing them in an electric field and observing their subsequent migration through a medium such as a gel **12** Angles of _________: the angle a beam radiation makes with a line perpendicular to the surface **14** Technique for separation of mixtures due to differing interactions with different media	**1** ____________ interference: two waves opposing each other and negating the effect of each other **2** Waves spread when passing through a crystal **5** _______ exchange resin: a polymer that acts as a stationary phase by attracting positively charged ions present in a solution **6** _______ solution: a solution than maintains a constant pH when small amount of acid and base are added **7** The ion formed when the molecule is ionised without being fragmented in a mass spectrometer **10** ______ exchange resin: a polymer that acts as a stationary phase by attracting negatively charged ions present in a solution **11** __________ phase: a solid to which the sample molecules adsorb as they move through the chromatography column **13** ________ field: a region of influence around a charged particle where a force is exerted on other charged objects **15** ______ phase: a liquid or gas that carries sample through a chromatography column **16** Stick to the surface

2 Write a short summary or create a mind map using all of the terms from the crossword.

19.2 IR spectroscopy and bond vibrations

Absorption of infrared radiation of a particular wavelength (and photon energy) gives rise to 'jumps' of the energy of vibrations of particular types in molecules. The energy of a vibrational mode between two atoms can have only definite values, and the only radiation that is absorbed is that which takes the vibrational energy to the next 'allowed' level. The gaps between the energies of vibration correspond with the energy of photons of infrared light.

For particular modes of vibration (e.g. stretching or bending) involving particular kinds of atoms in a molecule, the energy gap is similar from molecule to molecule, and so absorption of infrared radiation occurs at similar wavelengths. This means that in different substances, the carboxyl functional group, for example, absorbs infrared radiation of similar wavelengths. If there is absorption in the characteristic region for a carboxyl group, chemists can conclude that a substance has a carboxyl group in its molecules.

Which absorptions 'excite' the vibrations of which vibrational modes between which atoms? A program developed at King's College University in Edmonton, Canada is designed to show us exactly this. Go to the website: http://kcvs.ca/concrete/visualizations/chemistry

Scroll down to choose the applet *Functional groups, IR spectra and molecular vibrations.*

Click on the *HTML 5* icon to get the latest interactive version.

Select a functional group and then a compound. Its molecular structure will appear. Click on *Go to the IR spectrum* and the spectrum for that compound appears. Click on a peak in the spectrum and see which vibrational modes in the molecule are excited by absorption of radiation of that energy.

1 **a** Identify some compounds in which excitation of the same vibrational mode causes absorption of radiation of similar energy.

____

____

b What is the frequency of radiation that is absorbed (i.e. at what frequency is the absorption peak)?

____

____

c What is the vibrational mode (which type of vibration in which functional group) that gives rise to absorption at this frequency?

____

____

19.3 Isotopes and mass spectrometry

The relative atomic mass of a monoatomic element is a weighted average of the relative masses of each of its isotopes. A mass spectrometer does not measure the relative atomic mass of a monoatomic element: rather it measures the relative masses of its isotopes.

What happens when elemental molecular chlorine gas, Cl_2, is passed into a mass spectrometer? Ignore for now that some fragmentation (bond breaking) occurs so that there are peaks at the same places, and with the same relative abundance, for $^{35}Cl^-$ ions and $^{37}Cl^-$ ions, as shown in Figure 19.3.1.

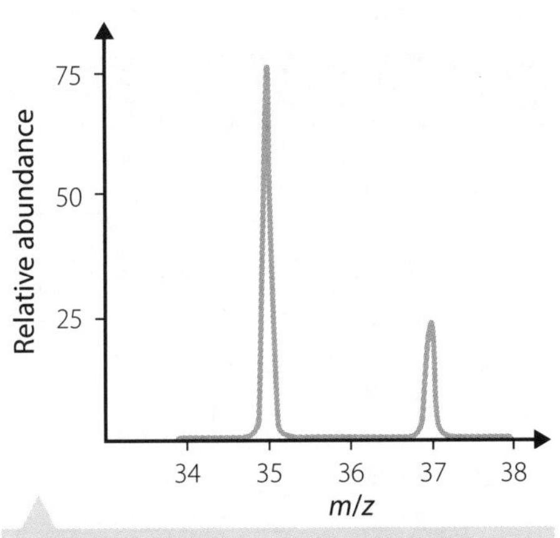

FIGURE 19.3.1 Mass spectra of different isotopes of chlorine gas

Some diatomic molecular ions are also formed and detected. What are their relative masses? In each Cl_2 molecule, the possible combinations of isotopes are:

^{35}Cl-^{35}Cl	with relative mass 70
^{35}Cl-^{37}Cl	with relative mass 72
^{37}Cl-^{37}Cl	with relative mass 74

1 Using the natural abundances of each of the isotopes of chlorine, estimate the relative heights of the mass spectral peaks at the following relative masses.

a 70 ______________________

b 72 ______________________

c 74 ______________________

2 Does your estimate correspond with the mass spectrum shown in the following diagram (which does not include peaks from fragment ions at relative masses 35 and 37)?

____

____

____

____

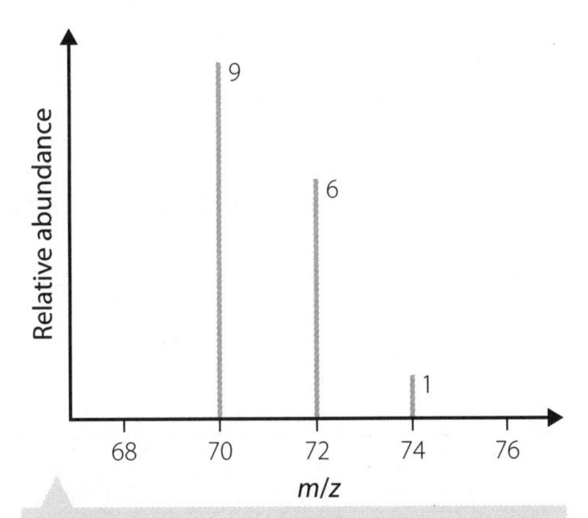

FIGURE 19.3.2 Mass spectra of different isotopes of chlorine gas

So far, we have been using integral values of the relative masses of isotopes. High-resolution mass spectrometry can measure relative isotopic masses very accurately. Accepted values for the most significant isotopes of carbon, hydrogen, oxygen, nitrogen and chlorine are given in the Table 19.3.1.

TABLE 19.3.1 Values for significant isotopes of carbon, hydrogen, oxygen, nitrogen and chlorine

ELEMENT	ISOTOPES	RELATIVE ABUNDANCE	RELATIVE MASS
Carbon	carbon–12 carbon–13	98.90% 1.10%	12.00000 13.00335
Hydrogen	hydrogen–1 hydrogen–2	99.99% 0.01%	1.00783 2.01410
Oxygen	oxygen–16 oxygen–18	99.76% 0.20%	15.99491 17.99916
Nitrogen	nitrogen–14 nitrogen–15	99.63% 0.37%	14.00307 15.00011
Chlorine	chlorine–35 chlorine–37	75.78% 24.20%	34.968852 36.965902

3 What are the possible combinations of isotopes in molecules of carbon monoxide (CO)? Which molecules are the most abundant, and what is their relative mass?

4 An analyst has a sample of a gas, which could be any one of carbon monoxide (CO), nitrogen (N_2), or ethylene (C_2H_4). High-resolution mass spectrometry shows that the most intense peak is at relative molecular mass 28.00614.

a What are the most abundant isotopic combinations in molecules of CO, N_2 and C_2H_4?

b What are the integral relative masses of the molecules with the most abundant isotopic combinations?

c What are the 'exact' relative masses of the molecules with the most abundant isotopic combinations?

d What is the gas sample?

9780170412476

1 The table below lists five amino acids, together with their molar masses, isoelectric points and R_f values when separated by paper chromatography using solvent S.

AMINO ACID	MOLAR MASS ($gmol^{-1}$)	ISOELECTRIC POINT	R_f VALUES
Arginine	174	10.76	0.20
Cysteine	121	5.02	0.40
Glutamic acid	147	3.08	0.30
Proline	115	6.30	0.43
Valine	117	6.02	0.61

A small sample of a mixture of some of these amino acids was placed at the top of a chromatography column and eluted downwards. Figure 19.4.1 shows the progress of the amino acids through the column after a period of time had elapsed.

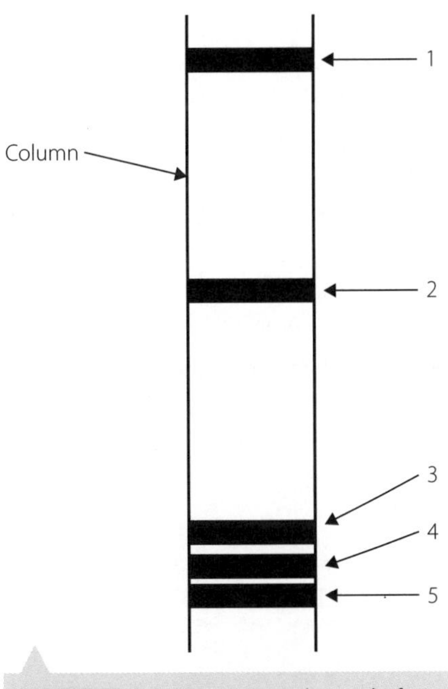

FIGURE 19.4.1 Chromatography results for five amino acids

a Explain why the amino acids separate into the bands shown in Figure 19.4.1.

____

____

____

____

____

____

b Complete the table below to indicate which band number in Figure 19.4.1 relates to which amino acid.

BAND NUMBER	AMINO ACID
1	
2	
3	
4	
5	

2 Separation and identification of the amino acids in a sample may also be made using gel electrophoresis. In this technique, charged substances are separated according to the different rates at which they move through a polyacrylamide gel when a potential difference is applied. The gel is moistened with a buffer solution so that the gel remains at a particular pH. At a particular pH some amino acids will be positively charged, some negatively charged and some will be 'zwitterions', ions that carry both a positive and a negative charge. The isoelectric point of an amino acid is the pH at which its zwitterion forms. A mixture of three amino acids, arginine, glutamic acid and valine, is separated by gel electrophoresis in a buffer solution at pH 6.0.

a Figure 19.4.2 shows the molecular structures for the three amino acids in the mixture. Draw the structure, showing all bonds, of valine in its zwitterion state.

Valine Arginine Glutamic acid

FIGURE 19.4.2 Separating a mixture: three amino acids

9780170412476

b Figure 19.4.3 shows the results of the gel electrophoresis. Complete the diagram to show a possible finished gel after the three amino acids have been separated, clearly labelling the locations of the glutamic acid and arginine amino acids.

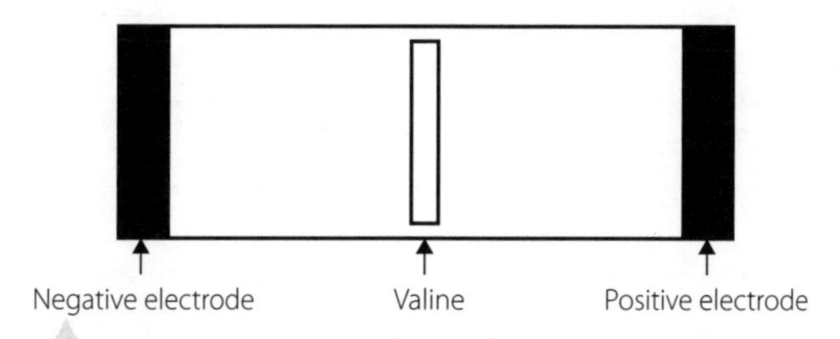

FIGURE 19.4.3 Separating a mixture: the gel electrophoresis results

3 Figure 19.4.4 shows the molecules of ethanol and dimethyl ether, both have the same molecular formula of C_2H_6O but different structural formulas. Two techniques, IR spectroscopy and mass spectrometry can be used to distinguish these molecules.

```
    H         H             H    H
    |         |             |    |
H — C — O — C — H       H — C — C — O — H
    |         |             |    |
    H         H             H    H
```

Dimethyl ether   Ethanol

FIGURE 19.4.4 Molecular structures of dimethyl ether and ethanol

a Explain the similarities and differences that would be observed in the infrared absorption spectra of ethanol and dimethyl ether, and thus how this technique could be used to distinguish the two molecules.

____

____

____

____

____

____

____

____

____

____

b Explain the similarities and differences that would be observed in the mass spectra of ethanol and dimethyl ether, and thus how this technique could be used to distinguish between the two molecules.

4 A mass spectrum of glycerol showed significant peaks at $m/z = 17, 31, 61$ and 92. Explain the existence of each of these peaks.

5 X-ray crystallography locates the positions of the nuclei of atoms or ions in a crystal. From this information, how can bond lengths and bond angles be determined?

9780170412476

6 Both acetone and prop-2-en-1-ol have the formula C_3H_6O. Their molecular structures are quite different, as shown in Figure 19.4.5.

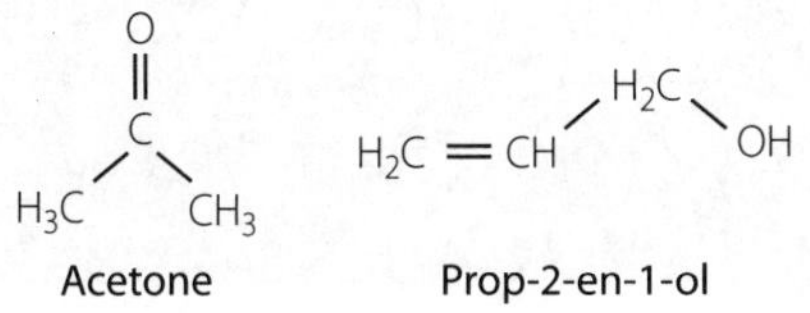

FIGURE 19.4.5 Molecular structures of acetone and prop-2-en-1-ol

How could you distinguish between samples of these two isomers by using IR spectroscopy?

___

___

___

___

___

___

20 Chemical synthesis

Summary

- Synthesis involves the use of chemistry to manufacture a new substance that has specific properties. Understanding the reaction mechanism in detail ensures that any chemical processes can take place as efficiently as possible, from both an economic and an environmental perspective.
- The yield of a reaction is the percentage of the reactant molecules that are converted to the desired products. When two chemicals react to produce a product, the chemical that is completely used up first is known as the limiting reagent. The other chemical that remains is known as the excess reagent. Therefore, the percentage yield is calculated by establishing the theoretical amount of product that would be formed assuming the limiting reagent is fully used up, and using this value in the following equation:

$$\%\text{ yield} = \frac{\text{actual mass of product formed}}{\text{theoretical mass of product formed}} \times 100$$

- Reaction conditions can be manipulated to maximise the yield by considering the appropriate position of equilibrium and also to ensure that the reaction is being carried out at an optimum rate.
- The Haber process is the industrial production of ammonia from gaseous nitrogen and hydrogen:

$$N_2(g) + 3H_2(g) \rightleftharpoons 2NH_3(g)\ \Delta H = -92.4\text{ kJ mol}^{-1}$$

  In order to maximise the production of ammonia the following method is used.
  - A compromise temperature of 400°C is used. This is low enough to produce an acceptable yield, given the reaction is exothermic, but also high enough to ensure an acceptable rate of reaction.
  - A compromise pressure of 250 atm is used. This is high enough to ensure that the equilibrium position lies to the right, given the volume of gas decreases as the reaction proceeds, but low enough to ensure an acceptable equipment cost.
  - A catalyst of iron or iron oxide is used to ensure that a high rate of reaction can still be achieved at a lower temperature.
  - The gases are mixed at a 3:1 ratio of $H_2:N_2$ as per the reaction equation.
- The Contact process is the industrial production of sulfuric acid. This occurs in a series of stages, the most significant of which is the production of sulfur trioxide from sulfur dioxide and oxygen:

$$2SO_2(g) + O_2(g) \rightleftharpoons 2SO_3(g)\ \Delta H = -196\text{ kJ mol}^{-1}$$

9780170412476

In order to maximise the production of sulfur trioxide in this reaction the following method is used.

- Gases are mixed to ensure that there is an excess of O_2 gas, to encourage the forward reaction.
- A compromise temperature of 400–450°C is used, low enough to favour the forward reaction as the reaction is exothermic, but high enough to ensure a reasonable rate of reaction.
- A pressure of 1–2 atm is used, low enough to reduce costs, but high enough to favour the forward reaction, as there is a decrease in volume from reactants to products.
- A catalyst is used to ensure that a high rate of reaction is achieved at a lower temperature.

- The following fuels can also be synthesised.
  - Ethanol, which can be obtained by fermenting sugars obtained from plant material:

    $$C_6H_{12}O_6(aq) \rightarrow 2C_2H_5OH(aq) + 2CO_2(g)$$

    This process is catalysed by yeast and is used in the production of alcoholic beverages, but also to produce ethanol as a fuel, as the ethanol can be extracted by distillation.
    Ethanol can also be obtained from the reaction of ethene with steam:

    $$CH_2 = CH_2(g) + H_2O(g) \rightarrow CH_3CH_2OH(l)\ \Delta H = -45\,\text{kJ mol}^{-1}$$

    This reaction is carried out at a low temperature and high pressure, with a catalyst of phosphoric acid in order to favour a high yield and high rate of production of ethanol.
  - Biodiesel is carried out through a process of transesterification. It is made from lipids that are naturally occurring in the form of vegetable oils. The lipid, a triglyceride ester, is heated with sodium hydroxide, methanol and a catalyst. This has the effect of hydrolysing the triglyceride ester, to form glycerol and a methyl ester with the resultant fatty acid. The methyl ester of the fatty acid can then be used to fuel diesel engines.
  - Hydrogen gas can be formed from the electrolysis of water, often using solar power, and then reacted in a fuel cell to generate electricity. A fuel cell is a galvanic cell but has a continuous supply of fuels and so a continuous production of electricity, so long as the fuel supply is maintained. The hydrogen gas is reacted with oxygen to produce water:

    $$2H_2(g) + O_2(g) \rightarrow 2H_2O(l)$$

    The fuel cell enables the oxidation of hydrogen and the reduction of oxygen to occur at different electrodes, the anode and cathode respectively, which are separated by a membrane to allow the passage of ions between the two electrodes. Different fuel cells can operate with either acid or alkaline electrolytes, depending on the preferred conditions.

20.1 Important terms

1 Complete the crossword using the clues on page 181.

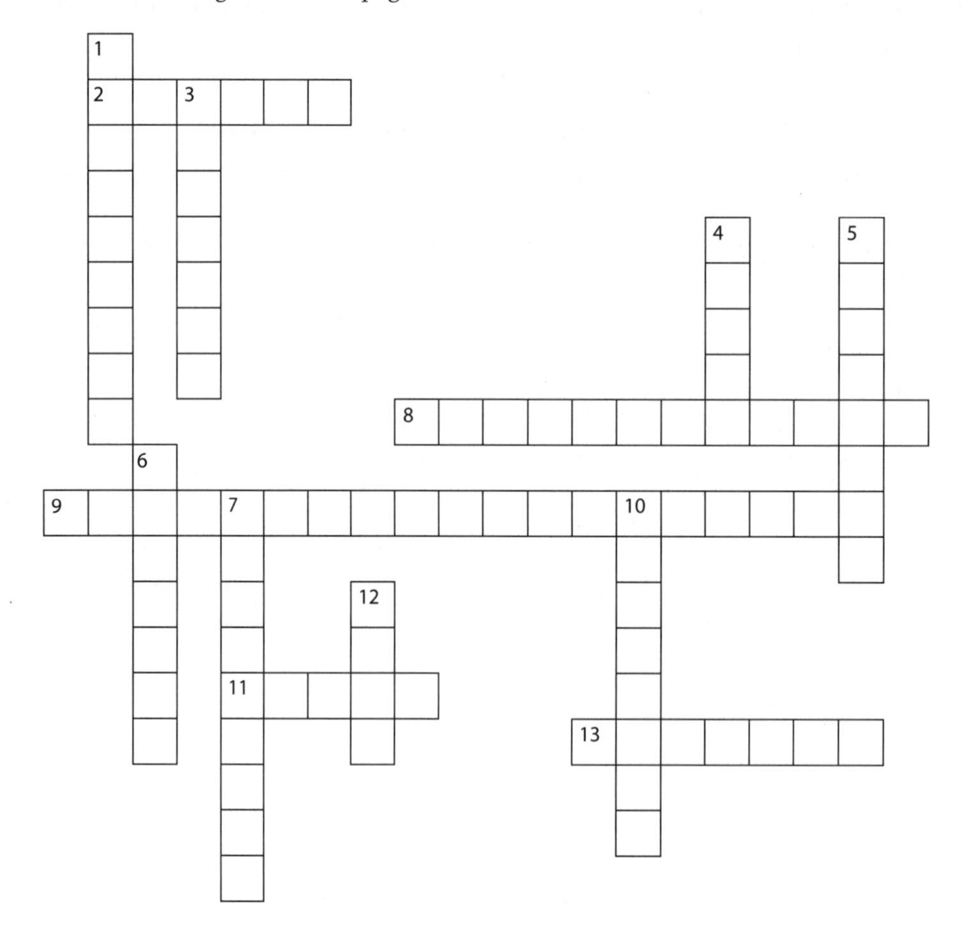

Across	Down
2 _______ reagent: the reactant of which some is still remaining after a reaction is complete **8** Species formed during a reaction pathway as a precursor to the desired product **9** _______: the process of converting one ester into another **11** _______ process: the industrial production of ammonia **13** A fuel produced from plant matter	**1** Reaction _________: an understanding and description of how a reaction has taken place **3** _________ process: the industrial production of sulfuric acid **4** Percentage _______: the percentage of the maximum possible amount of product from a reaction that has actually been produced **5** _________ reagent: the reactant molecule that is completely used up during a reaction **6** Reaction ________: a series of reactions leading to a specific outcome or product **7** Created artificially rather than naturally occurring **10** Catalytic _________: the process of heating larger organic molecules and converting them into smaller, more useful molecules **12** _____ cell: a galvanic cell that generates electricity from a chemical reaction, but where there is a continuous supply of fuel and also removal of products

2 Write a short summary or create a mind map using all of the terms from the crossword.

20.2 Theoretical and percentage yield

Theoretical yield is the maximum amount of a product that could be formed in a reaction mixture, given the initial amounts of reactant present, and the stoichiometry of the reaction (as indicated by the balanced chemical equation).

1 Consider the precipitation reaction represented by the following balanced chemical equation:

$$Ca^{2+}(aq) + 2F^{-}(aq) \rightarrow CaF_2(s)$$

Suppose that an aqueous solution containing 0.0040 mol of $Ca^{2+}(aq)$ ions, such as a calcium nitrate solution, is mixed with another solution containing 0.0020 mol of $F^{-}(aq)$ ions, such as a potassium fluoride solution.

a What is the limiting reactant?

b What is the theoretical yield of $CaF_2(s)$ precipitate (in moles)?

The percentage yield refers to the mass of a product that is produced, as a percentage of its theoretical yield. A very simple analogy is that if 100 beans are planted in soil, the theoretical yield of bean plants is 100. If only 87 seeds germinate to become plants, the percentage yield is 87%.

2 Methanol can react in the presence of an appropriate catalyst to form hydrogen gas and carbon monoxide gas. The hydrogen can be used as a fuel, or these products can be used to make other substances:

$$CH_3OH(l) \rightarrow CO(g) + 2H_2(g)$$

a If 1000 g of methanol reacts in this way, what is the theoretical yield (mass) of hydrogen gas?

9780170412476

b If 103.5 g of hydrogen is produced, what is the percentage yield?

20.3 Raw materials

1 What are the raw materials used in the production of ammonia by the Haber process; in other words, what are the naturally occurring substances that are used in one or more reactions to produce the desired product?

2 Why is this process said to give us the ability to turn 'air into bread'?

3 Why is the development of this process so remarkable? What has it made possible that was previously not possible?

4 What were the challenges facing Haber and others in developing this process?

5 Search information sources to find out about other aspects of the chemistry career of Fritz Haber. Do you think he was a great person?

20.4 Extent of reaction and temperature

In an experiment conducted by Fritz Haber, nitrogen gas was introduced into a reaction chamber at a concentration of 0.00500 mol L^{-1}, and hydrogen at 0.01250 mol L^{-1}. With the reaction mixture at 472°C, a reaction to form ammonia took place and when equilibrium was reached, the concentration of ammonia was found to be 3.18×10^{-5} mol L^{-1}.

1 What is the value of the equilibrium constant for the Haber process reaction at 472°C, represented by the following equation?

$$N_2(g) + 3H_2(g) \rightleftharpoons 2NH_3(g)$$

9780170412476

Complete the calculation in the following steps. Consider a 1.00 L sample of the reaction mixture.

a What were the initial amounts (in mol) of $N_2(g)$ and $H_2(g)$ admitted to the chamber?

b What was the amount (mol) of $NH_3(g)$ present at equilibrium?

c **i** Using the balanced equation, how much (in mol) of $N_2(g)$ had reacted before equilibrium was reached?

ii What is the residual equilibrium amount (in mol) of $N_2(g)$?

iii What is the equilibrium concentration of $N_2(g)$?

d **i** Using the balanced equation, how much (in mol) of $H_2(g)$ had reacted before equilibrium was reached?

ii What is the residual equilibrium amount (in mol) of $H_2(g)$?

iii What is the equilibrium concentration of $H_2(g)$?

e Substitute the equilibrium concentrations of reactants and product in the expression for the equilibrium constant and deduce its mathematical value.

2 What percentage of the initial amount of $N_2(g)$ reacted before equilibrium was attained?

3 What percentage of the initial amount of $H_2(g)$ reacted before equilibrium was attained?

4 Would you expect the amount of $N_2(g)$ and $H_2(g)$ that react before equilibrium is attained to be greater, less or the same, if the reaction mixture had been at a lower temperature? Explain.

20.5 The role of catalysts in the production of materials from syngas

Synthesis gas, also called syngas, is essentially a mixture of carbon monoxide and hydrogen. Syngas can be produced from many raw materials, such as natural gas, coal, biomass or hydrocarbon feedstocks, like petroleum, by reaction with steam or oxygen. The production of syngas is described as partial oxidation, because the complete oxidation of organic materials would result in formation of carbon dioxide. Examples of reactions to produce syngas include the following.

Reaction of natural gas with steam:

$$CH_4(g) + H_2O(g) \rightarrow CO(g) + 3H_2(g)$$

Reaction of coke with steam:

$$C(s) + H_2O(g) \rightarrow CO(g) + H_2(g)$$

Syngas from these two sources, for example, differs in the ratio of carbon monoxide and hydrogen. Syngas is an extremely important industrial intermediate resource for production of hydrogen, ammonia, methanol and synthetic hydrocarbon fuels similar to petrol or diesel.

1 Why do you think that in some places hydrocarbon fuels are used to make syngas for production of other materials (like methanol), while in other places coal is used to make syngas for production of synthetic hydrocarbon fuels?

2 Each of these reactions requires different bond-breaking and bond-making processes. Different catalysts have been identified, or designed, by chemists to change syngas to particular products.

Consult sources, including Internet websites, to find which catalysts are used for each of the following purposes.

a Conversion of syngas, by reaction with steam into a mixture of hydrogen and carbon dioxide, from which the carbon dioxide is easily removed by cooling, so that nearly pure hydrogen can be used as a fuel, or as a feedstock in the Haber process to make ammonia. This is called a 'shift' reaction because it can be used to alter the ratio of hydrogen and carbon monoxide in syngas:

$$CO(g) + H_2O(g) \rightarrow CO_2(g) + H_2(g)$$

b Conversion of syngas into methanol:

$$CO(g) + 2H_2(g) \rightarrow CH_3OH(l)$$

9780170412476

c Conversion of syngas into synthetic hydrocarbon fuels, called the Fischer–Tropsch synthesis:

$$nCO(g) + (2n + 1)H_2(g) \rightarrow C_nH_{2n} + 2(l) + nH_2O(g)$$

$$nCO(g) + 2nH_2(g) \rightarrow C_nH_{2n}(l) + nH_2O(g)$$

20.6 Revision questions

1 List five of the most important naturally occurring raw materials that are used in the manufacture of chemical products, either directly or from secondary feedstocks. In each case, list some of the important products obtained from these raw materials (or from substances made from the raw materials).

2 What factors govern the feasibility of an industrial process being considered for synthesis of a particular product?

3 List the characteristics of an ideal synthetic process.

4 a What challenges are presented by the Haber process for manufacture of ammonia?

b How are these challenges related to the fact that design of the process was itself a challenge?

9780170412476

c How do manufacturers adjust the reaction conditions to deal with these challenges, in order to maximise yields and save on energy costs?

5 What compromises in economy of materials or energy are necessary in the production of sulfuric acid by the contact process?

6 List the ways that the yield of a synthetic process can be maximised.

7 Distinguish 'theoretical yield' and 'percentage yield' of a process, whether in industry or in the laboratory.

8 Give at least one example that demonstrates the specificity of action of a catalyst.

9 It is often claimed that petroleum products should not be used as fuels, because they are too important as sources of chemical products. What do you think?

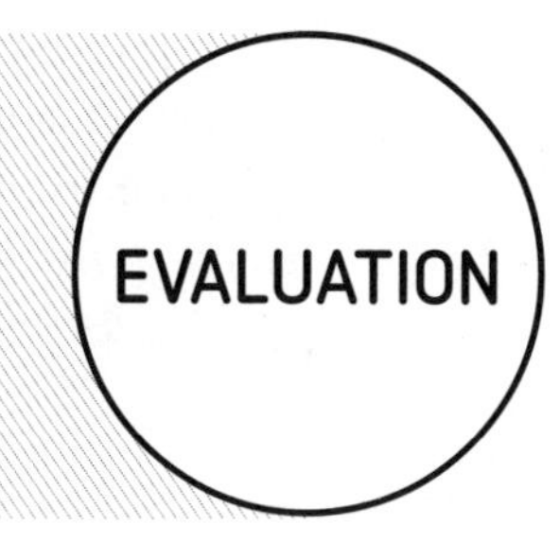

1 Ethanol can be produced industrially by reacting ethene with steam according to the equation:

$$CH_2CH_2(g) + H_2O(g) \rightleftharpoons CH_3CH_2OH(g) \ \Delta H = -45 \text{ kJ.mol}^{-1}$$

The reaction conditions are:

- $T = 300°C$
- $P = 60–70$ atm
- catalyst = phosphoric(V) acid.

a A conflict is involved in choosing the best temperature for this reaction. Suggest a reason for this conflict and explain how it is resolved.

b If a higher pressure were used there would be a greater equilibrium yield of ethanol. Explain why this is the case and then why a higher pressure is not in fact used in this reaction.

2 Biodiesel is produced by transesterification, a process in which triglycerides are converted to methyl esters in a reaction catalysed by either an acid or a base. Research has shown that cooking oil (e.g. the vegetable oil used for frying chips) is suitable for the production of biodiesel. The transesterification of a triglyceride in cooking oil is represented by the incomplete equation below.

$$\begin{array}{l} CH_2-O-\overset{\overset{\displaystyle O}{\|}}{C}-C_{17}H_{33}- \\ | \\ CH-O-\overset{\overset{\displaystyle O}{\|}}{C}-C_{15}H_{31} \\ | \\ CH_2-O-\overset{\overset{\displaystyle O}{\|}}{C}-C_{17}H_{31} \end{array} + 3\mathbf{A} \longrightarrow \begin{array}{l} C_{17}H_{33}COOCH_3 \\ C_{15}H_{31}COOCH_3 \\ C_{17}H_{31}COOCH_3 \end{array} + \mathbf{B}$$

a Write the name and give the semi-structural formula of compound **A**.

____

b Give the name and molecular formula of compound **B**.

____

c Give the semi-structural formula of the biodiesel produced that is monounsaturated.

____

____

21 Green chemistry

Summary

- Green chemistry involves reducing the impact of chemistry on the environment by reducing or eliminating the use or production of hazardous substances.
- The principles of green chemistry are as follows.
  1. Prevent waste
  2. Maximise atom economy
  3. Design less hazardous chemical syntheses
  4. Design safer chemicals and products
  5. Use safer solvents and reaction conditions
  6. Increase energy efficiency
  7. Use renewable reactants
  8. Avoid using chemical derivatives
  9. Use catalysts
  10. Design chemicals that will degrade
  11. Prevent pollution
  12. Minimise the potential of chemical accidents
- Atom economy is a measure of the percentage of the atoms from the reactants that are contained in the desired product, and hence are not 'wasted' or required to be disposed of in any other way as part of the chemical process. Therefore, the higher the atom economy, the greener the process.

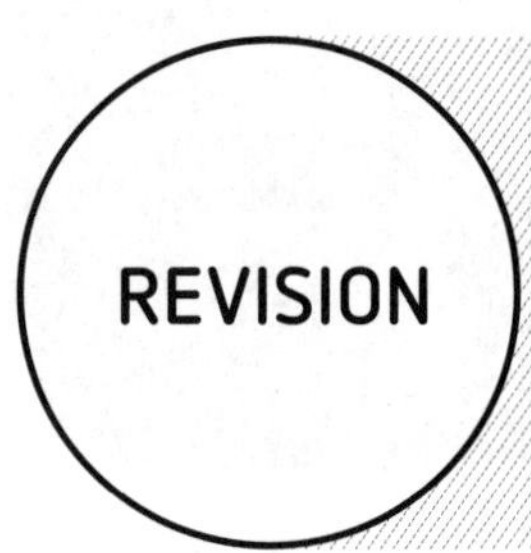

21.1 Important terms

1 Complete the crossword using the clues below.

Across	**Down**
2 Atom ___________: the percentage of atoms in the reactants that are in the desired product	**1** The escape or release of chemicals that negatively affect the local environment
3 A chemical obtained as the result of a chemical process rather than being produced naturally	**4** A chemical that causes a potential risk
7 _____ chemistry: the reduction of the impact of chemistry on the environment	**5** A chemical that is used to dissolve another
8 Energy ___________: a reaction that occurs with the lowest possible loss of energy to the surroundings	**6** A form of energy or chemical of which there is an inexhaustible supply
9 A chemical that increases the rate of a reaction without itself being used up in the reaction	

9780170412476

2 Write a short summary or create a mind map using all of the terms from the crossword.

21.2 Calculating atom economy

Researchers are currently exploring alternative methods for producing methanol, in an attempt to increase the atom economy. The most important current method is to make syngas from methane and steam, and then to react components of the syngas in the presence of a copper/zinc catalyst, as seen in the following reactions:

$$CH_4(g) + H_2O(g) \rightarrow CO(g) + 3H_2(g)$$

$$CO(g) + 2H_2(g) \rightarrow CH_3OH(g)$$

The overall stoichiometry of these two steps (achieved by adding the two equations) is:

$$CH_4(g) + H_2O(g) \rightarrow CH_3OH(g) + H_2(g)$$

1 **a** Calculate the atom economy of the production of methanol using this method.

b What other factors govern the desirability of producing methanol in this way?

c If the excess $H_2(g)$ can be used for some other purpose, so that both products are desirable, what is the atom economy of the reaction?

Professor George Olah and co-researchers at the University of Southern California are investigating ways to make methanol that are more atom-efficient as well as more energy-efficient. One idea under consideration is a direct reaction of methane and oxygen in the presence of a catalyst that produces different products from the combustion of methane. Olah said, 'You take a methane molecule and stick in just one oxygen atom.'

$$2CH_4(g) + O_2(g) \rightarrow 2CH_3OH(g)$$

2 What is the atom economy of production of methanol in this way?

However, there is a problem. Just because we can write a chemical equation does not mean that the reaction that it represents will proceed. No catalyst has yet been developed that can bring about this reaction at moderate temperatures.

Another method being investigated depends on the development of catalysts that allow methane and bromine to react at moderate temperatures to produce bromoethane, which can then be passed into water where it reacts to form methanol:

$$CH_4(g) + Br_2(g) \rightarrow CH_3Br(g) + HBr(g)$$

$$CH_3Br(g) + H_2O(g) \rightarrow CH_3OH(g) + HBr(g)$$

The overall stoichiometry associated with these two reactions is found by adding the equations:

$$CH_4(g) + Br_2(g) + H_2O(g) \rightarrow CH_3OH(g) + 2HBr(g)$$

3 Calculate the atom economy of the production of methanol in this way.

The atom economy of this last method seems low because the Br atoms do not finish up in the product. However, if the HBr is separated from the system, and then allowed to react with air, bromine (Br_2) is regenerated, and this can be recycled to the reaction with methane. If HBr is not a waste product to be disposed of, the atom efficiency of this (currently not operational) method rises to the same as that for production of methanol from methane and steam, and the energy costs would be much less.

The message here is that, while we should not waste atoms, sometimes it is misleading to simply calculate the atom economy of an isolated reaction without considering the uses of all of the products.

21.3 The manufacture of ibuprofen

Ibuprofen is one of the major painkilling medicines sold over the counter in pharmacies across the world. It was first developed in 1961 and patented by the Boots company. It is now available as a generic medicine across the world.

The structure of ibuprofen is shown in Figure 21.3.1.

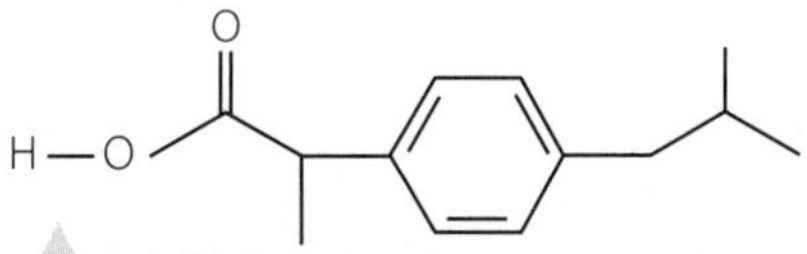

FIGURE 21.3.1 Ibuprofen

1 What are the functional groups present in an ibuprofen molecule?

____

The initial process of the manufacture of ibuprofen developed by the company Boots involved a six-step procedure, which is shown in Figure 21.3.2. The raw material used was the molecule 2-methylpropylbenzene.

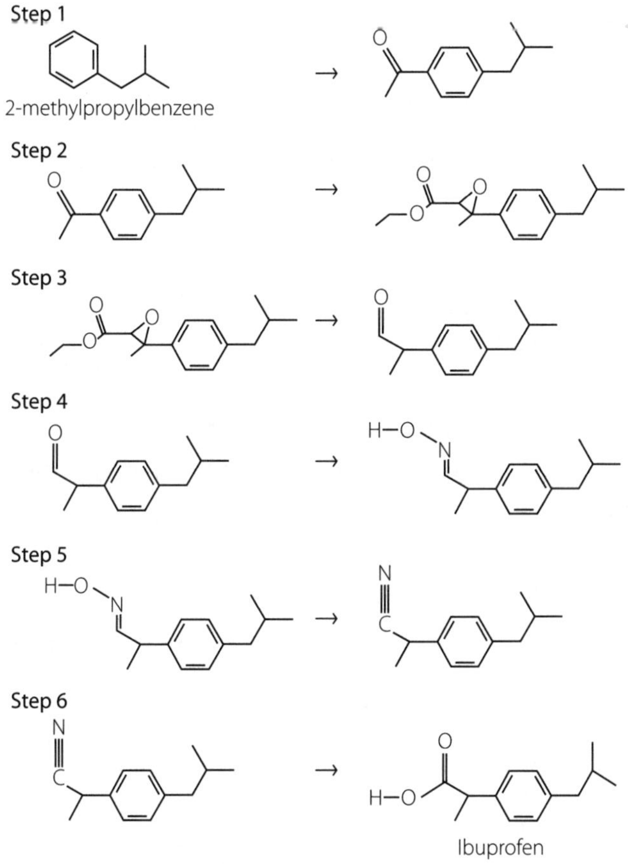

FIGURE 21.3.2 Six-step manufacture of ibuprofen

2 Typically, a very efficient chemical reaction can be carried out with a percentage yield of 80%. Assuming that each step in the manufacture of ibuprofen has an 80% yield, calculate the overall percentage yield of ibuprofen from 2-methylpropylbenzene.

____

____

____

____

3 The atom economy for this manufacture, taking into account the additional reagents that are used in each step, comes to 40%. If 1500 tonnes of ibuprofen is consumed each year in Australia, calculate the total mass of waste product also produced.

____

____

____

____

An alternative process for the manufacture of ibuprofen has since been developed, which involves only three steps. It is shown in Figure 21.3.3.

FIGURE 21.3.3 Three-step manufacture of ibuprofen

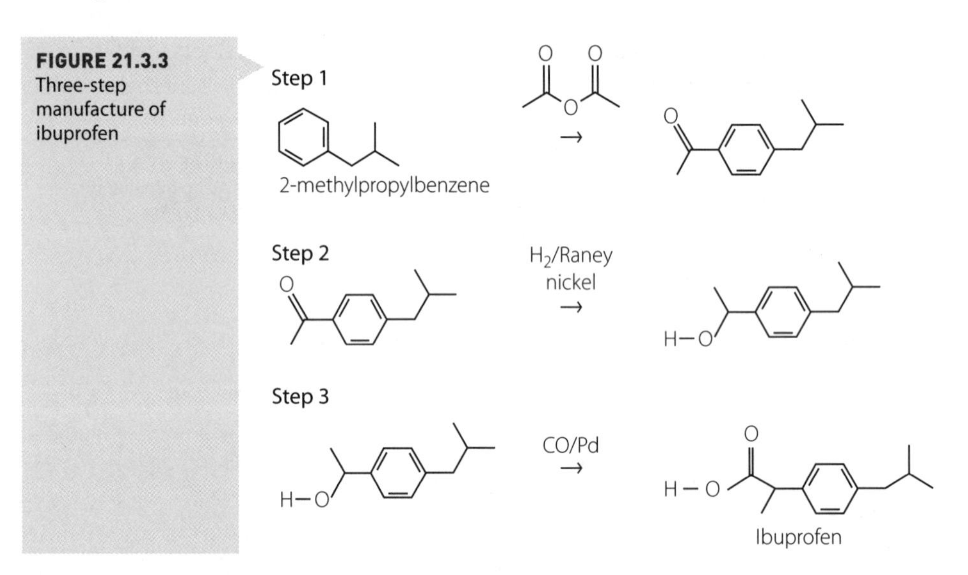

4 Assuming the same percentage yield of each step in the process, compare the overall yield of ibuprofen from this process compared to the original method.

____

____

____

____

5 The atom economy of the three-step process is 77%, double that of the original process. With particular reference to Step 2 in the process, identify some design features of this new process that have led to the significant improvements in yield and atom economy.

1 When developing a green chemical process a chemist would aim to:

A achieve the highest product yield possible.

B carry out the process using a continuous flow process.

C maximise the atom economy for the reactions involved.

D maximise the rate at which the product is formed.

2 Under certain conditions of temperature and pressure, carbon dioxide behaves as a supercritical fluid, and can adopt properties that are midway between a gas and a liquid. Consequently, supercritical carbon dioxide has a range of applications. One such application is its use to extract caffeine from, and therefore 'decaffeinate', coffee beans.

The process is as follows:

Step 1 The supercritical CO_2 is forced through the coffee beans

Step 2 Caffeine from the coffee is extracted by the supercritical CO_2

Step 3 High pressure is sprayed onto the CO_2 and dissolves the caffeine

Step 4 CO_2 gas is released from the solution

Step 5 Caffeine isolated and sold to drug companies

Step 6 Coffee beans are roasted to make decaffeinated coffee

a Propose an additional design feature to this process to ensure that there is no net increase in the concentration of atmospheric CO_2.

b Select three of the principles of green chemistry and outline how the process of decaffeination complies with these three principles.

22 Macromolecules: Polymers, proteins and carbohydrates

Summary

- Addition polymerisation involves the addition of many monomers together. The monomer must have a double bond that breaks during the polymerisation process, enabling each carbon to form a bond with another monomer.
- The simplest addition polymer is polyethene. Other addition polymers include polystyrene, polytetrafluoroethene, polyvinyl chloride and polymethyl methacrylate.
- The structure of the monomer can determine the properties of the polymer.
- Polymer chains can be closely packed together or loosely ordered.
- Condensation polymers are formed when two monomer molecules join together to eliminate a small molecule (often water).
- Examples of condensation polymers include polyesters such as dacron, and polyamides such as nylon.
- Polypeptides are in important class of naturally occurring condensation in which the monomer is an amino acid.
- Amino acids contain the amino (—NH_2) functional group and the carboxyl (—COOH) functional group.
- The link that joins two amino acids together is called a peptide link. The compound formed when two amino acids join together is called a dipeptide and when long chains of amino acid units join together the result is a polypeptide.
- Polysaccharides are an important class of naturally occurring condensation polymers. In the case of polysaccharides, the monomers are monosaccharides.
- Monosaccharides undergo condensation polymerisation to form polysaccharides.
- The link that joins two monosaccharides together is called a glycosidic bond. The compound formed when two monosaccharides join together is called a disaccharide and when long chains of monosaccharide units join, the result is a polysaccharide.

9780170412476

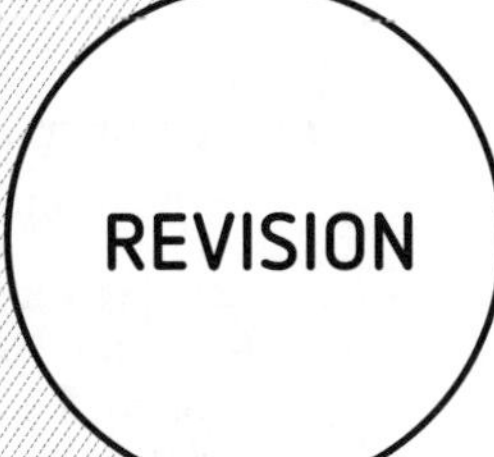

22.1 Addition polymerisation

Addition polymerisation uses monomers that contain double bonds. During the polymerisation process when, for example, ethene gas is placed under conditions of high temperature and pressure in the presence of a catalyst, one of the bonds in the double bond is broken.

This leaves carbon atoms available to form bonds with carbon atoms from neighbouring monomers. This process is shown in Figure 22.1.1.

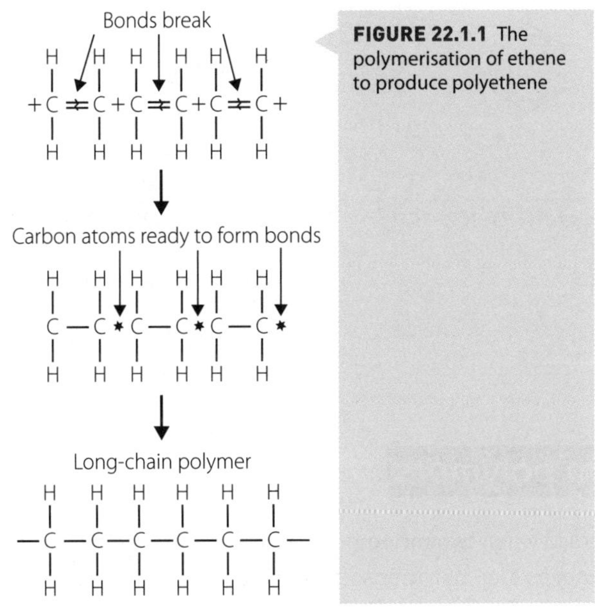

FIGURE 22.1.1 The polymerisation of ethene to produce polyethene

1 Propene can give rise to a number of useful forms of the polymer ____________________.

FIGURE 22.1.2 Propene

2 In the boxes below, sketch and name the three possible arrangements of the CH_3 groups in the polymer chain.

____________tactic	____________tactic	____________tactic

3 A number of addition monomers are given in the table below. For each one, sketch a polymer chain containing ten of the monomer units and give the name of the polymer.

MONOMER NAME	STRUCTURAL FORMULA	POLYMER
Chloroethene (vinyl chloride)	$CH_2{=}CHCl$	
Tetrafluoroethene	$F_2C{=}CF_2$	
Styrene	$CH_2{=}CH$ with a benzene ring attached to CH	
Vinyl acetate	$CH_2{=}CH{-}O{-}CO{-}CH_3$	

22.2 Condensation polymers

Condensation polymers are formed when two monomer molecules join together to eliminate a small molecule (often water). The process is called condensation polymerisation, and there are two groups of condensation polymers.

- Synthetic condensation polymers: polyesters and polyamides
- Natural condensation polymers: polysaccharides (cellulose, starch) and proteins

POLYESTERS

Polyesters can be formed either by a single monomer or a combination of monomers.

Single monomer: one that has a carboxylic acid functional group (—COOH) at one end and an alcohol group (—OH) at the other, such as 2-hydroxypropanoic acid (polylactic acid).

HO — CH(CH_3) — C(=O) — OH

Hydroxyl functional group

Carboxyl functional group

FIGURE 22.2.1 2-hydroxypropanoic acid

1 Figure 22.2.2 shows the structural formula of hydroxyethanoic acid (also known as hydroxyacetic acid, hydroacetic acid or glycolic acid).

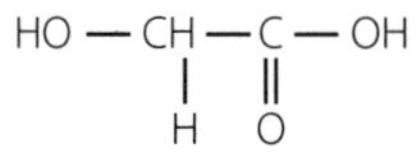

HO — CH — C — OH
       | ||
       H   O

FIGURE 22.2.2 Hydroxyethanoic acid

Draw the structure of polyglycolic acid containing five monomer units. On the diagram, clearly show where the water molecules are being eliminated from and identify the ester links.

Two different monomers: one monomer would be a diol (two —OH groups) and the other would be a dicarboxylic acid, as shown in Figure 22.2.3.

$HO-CH_2-CH_2-OH$

Ethylene glycol

HO — C(=O) — C_6H_4 — C(=O) — OH

Terephthalic acid

FIGURE 22.2.3 Ethylene glycol and terephthalic acid

2 Draw the structure of the polymer produced from these monomers (five repeating monomer units).

22.3 Carbohydrates

1 Below are two forms of glucose, α and β. Identify and label each type.

__glucose   __glucose

2 Sucrose (sugar) is formed when α-glucose and β-fructose undergo a condensation reaction.

$\rightarrow$ + H_2O

On the diagram in question 1, clearly show where the water molecule came from and identify the glycosidic link in the sucrose molecule.

3 Undertake research on the polysaccharides: starch, cellulose and glycogen.
Topics to search for include:

- function
- where they are found
- how they occur (e.g. grains? fibres? other?)
- from which monomers are they made
- how the polysaccharide is arranged (straight, branched or coiled chains).

Summarise your findings in the table below.

	CELLULOSE	STARCH	GLYCOGEN
Function			
Where they are formed			
How they occur			
Monomer			
How the polysaccharide is arranged			

22.4 Proteins

1 The structure of proteins can be divided into four levels: primary, secondary, tertiary and quarternary.

a Choose one of the sentences below as the most appropriate description of each structure and include it in the correct box.

The three-dimensional shape of a protein.

The way that amino acids in a protein chain bond to amino acids in the same or nearby chains.

The sequence of the amino acids in a protein chain.

The three-dimensional structure consisting of the aggregation of two or more individual polypeptide chains that operate as a single unit.

b Draw a diagram to show the shape of each structure.

Primary structure
Secondary structure
Tertiary structure
Quaternary structure

22.5 Amino acids

Amino acids contain the amino (—NH_2) functional group and the carboxyl (—COOH) functional group.

The simplest amino acid is glycine.

```
        H   O
        |   ||
  H—N—C—C—OH
     |   |
     H   H
```

1 Show how two glycine molecules undergo a condensation reaction to form the dipeptide GlyGly. Clearly show where the water comes from and identify the peptide link.

2 Draw the structure of the polypeptide AlaGlyAlaLys showing the peptide links (Hint: refer to Fig.18.2.5, p. 278 of the Student Book.).

1 The monomer with the structure:

$$\mathrm{H_2C{=}CH{-}O{-}C({=}O){-}CH_3}$$

forms which common polymer?

A PVC
B PVA
C polystyrene
D HDPE

2 A zwitterion is:

A a type of monomer used is condensation polymerisation.
B an ion that forms a positive and negative charge on the same group of atoms.
C an intermediate stage in the formation of a polysaccharide.
D an additive used to stabilise and strengthen addition polymers.

3 Vulcanisation is the process whereby:

A straight-chain polymers have side groups added to increase rigidity.
B increasing the crystallinity of a polymer to decrease permeability.
C increasing the chain length of a polymer in order to strengthen it.
D sulfur is added to a polymer to increase polymer strength.

4 A condensation polymer can be produced by combining:

A a diol and a dicarboxylic acid.
B a monosaccharide and a substituted alkene.
C substituted alkenes.
D an amino acid and a carboxylic acid.

5 Compare and contrast the properties and structures of LDPE and HDPE.

9780170412476

6 Explain why poly(vinyl chloride) forms a hard, rigid polymer yet polystyrene is hard but brittle, when they are both amorphous polymers.

7 Refer to your answer to question 1 in section 22.1. Suggest how these different arrangements might affect the structure and properties of the polymer.

23 Molecular manufacturing

Summary

- Molecular manufacturing is the production of chemicals with specific shapes or chemical composition.
- Molecular manufacturing can be done in three ways:
  - the orientation effect, also known as mechanosynthesis
  - manipulation of structures at the atomic or molecular level
  - the use of protecting groups.
- The *top-down* approach to molecular manufacturing involves converting molecules into a smaller, more useful substance. The problem with this method is that it can use vast amounts of energy and materials, and produce a lot of waste.
- The *bottom-up* approach involves building a desired structure from the atomic and molecular scale, by adding atoms individually or by adding molecules together to make a larger, more complex molecule. This method is more difficult because it is done on the atomic scale, but it has less waste and requires much less energy.
- Artificial protein synthesis has been carried out in bacterial and yeast cells. These have been used to produce diabetes medications, clotting drugs to treat haemophilia and human growth hormone to treat dwarfism.
- A carbon nanotube is formed from sheets of graphene, covalently bonded carbon, with each sheet only one atom thick. The sheets are rolled into long tubes, and can be single walled, double walled or have a variety of other shapes. Carbon nanotubes are the strongest and stiffest materials yet discovered or made. Carbon nanotubes have been used to develop strong but flexible materials for use as artificial muscle and bone implants in humans. One of the most useful properties of carbon nanotubes is their high electrical conductivity. This makes them useful in batteries and fuel cells to produce electricity.
- Nanorobots are artificially created proteins designed to perform specific mechanical tasks.
- A nanosensor is a sensor that is nanosized. For example, biosensors can detect when a particular chemical is present in the blood or body and indicate the presence of disease.

23.1 The orientation effect

1 Figure 23.1.1 shows an example of a successful collision between two molecules.

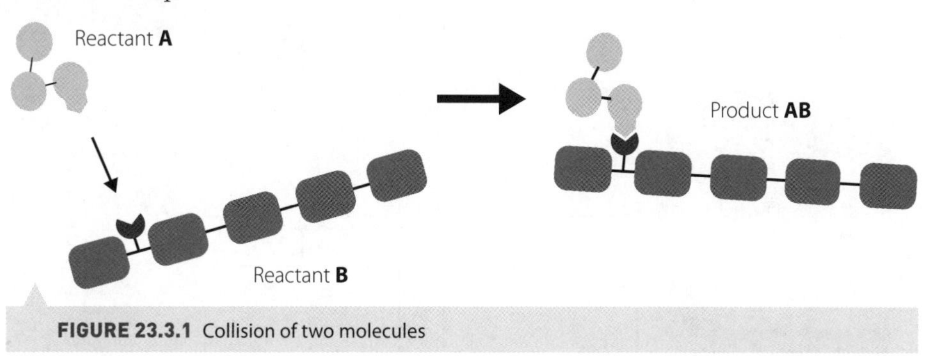

FIGURE 23.3.1 Collision of two molecules

a Redraw the sketch to show what would happen if the reacting molecules **A** and **B** collided unsuccessfully.

b How could you improve the chances of successful collisions occurring? Add this to Figure 23.3.1.

23.2 Top-down and bottom-up approaches

The following diagrams show the stages used in the manufacturing of a component in a microelectronic circuit. The finished component is a substrate with two electrodes connected by a conductor.

1 Match the diagram with the sequence by placing the appropriate letter into the correct numbered box to show both top-down and bottom-up approaches to fabricating the component. Place the correct letter in each box to create the top-down and bottom-up approaches to fabricating the component.

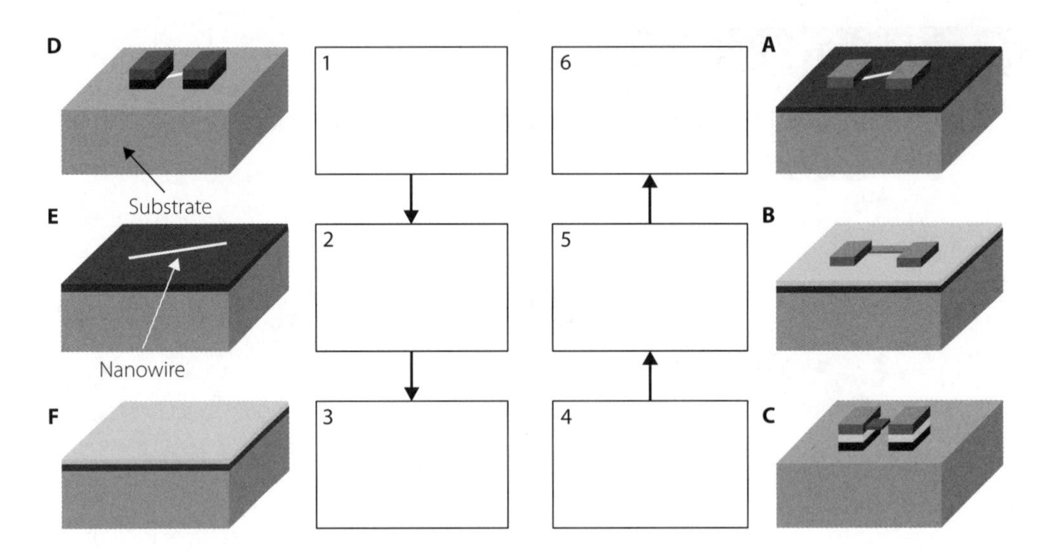

23.3 Synthetic proteins

Synthetic proteins can be extremely useful, for example, in finding applications to treating diseases such as diabetes and haemophilia, and manufacturing super-strong synthetic spider silk.

However, many of these applications require the rare D-form of the amino acid protein building blocks.

Amino acids have two forms: L-form and D-form. These are mirror images of each other, as shown in Figure 23.3.1.

FIGURE 23.3.1 L-form and D-form of amino acids

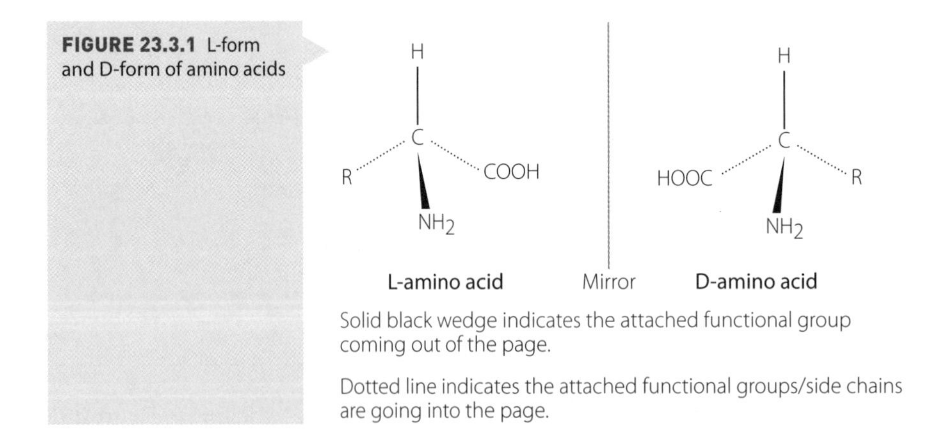

9780170412476

A protein has been developed to enhance the strength of cotton fibres. The sequence of the amino acids in the protein is very similar to that found in fingernails but with a few key differences. A part of the sequence is shown below.

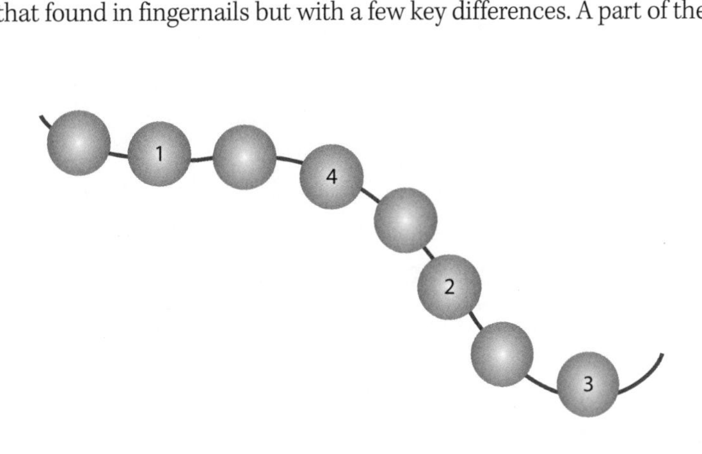

The sequence requires synthetic amino acids at points **1–4**, as shown in in the table.

TABLE 23.3.1 Identifying amino acids in a protein

NUMBER	NAME	STRUCTURE	IDENTIFYING LETTER
1	Valine	$CH_3-CH-CH_3$ | $H_2N-CH-COOH$	
2	Alanine	CH_3 | $H_2N-CH-COOH$	
3	Threonine	$CH_3-CH-OH$ | $H_2N-CH-COOH$	
4	Phenylalanine	$C_6H_5-CH_2$ | $H_2N-CH-COOH$	

Figure 23.3.2 shows both L- and D-forms of the amino acids in the table in their three-dimensional representations. It should be noted that the desired protein needs the amino acids required to be in their D-forms.

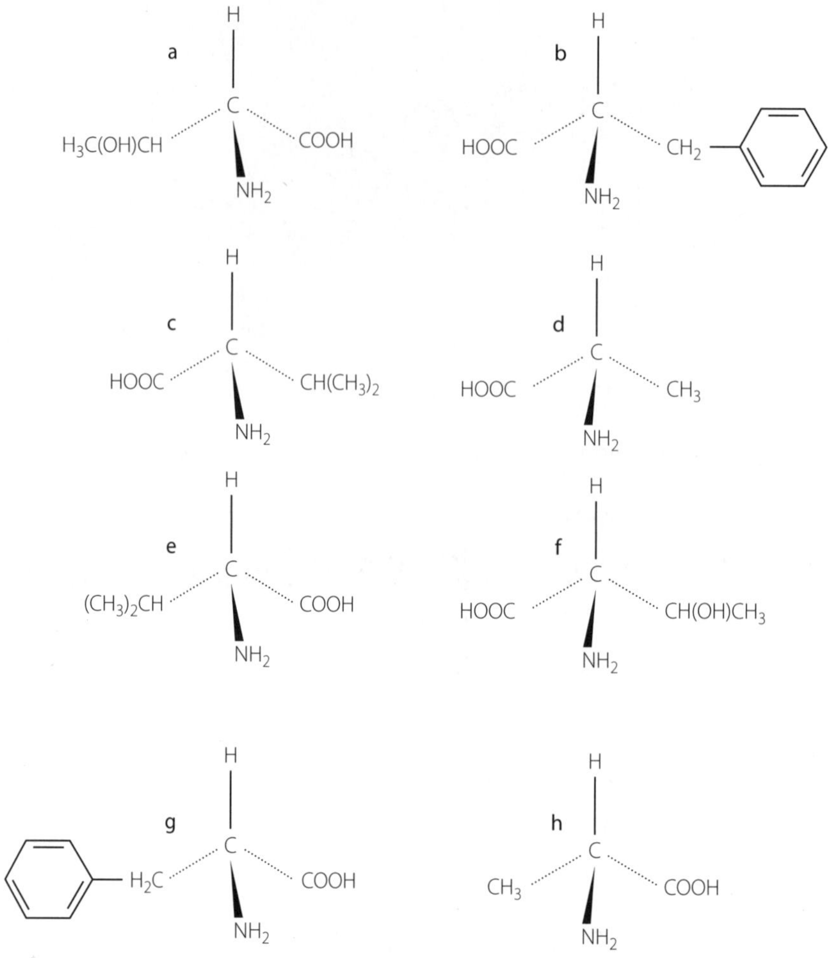

FIGURE 23.3.2 L- and D-forms of the amino acids in Table 23.3.1

1 Use the structures in Figure 23.3.2 to identify the amino acids and write the appropriate letter of the D-form of each amino acid in Table 23.3.1.

9780170412476

1 Another name for the orientation effect is:

A chemical mechanics.

B isometric effect.

C mechanosynthesis.

D collision theory.

2 The functional groups that must align in the formation of a dipeptide are:

A an amine group and a carboxyl group.

B two amine groups.

C an amine group and a hydroxyl group.

D two carboxyl groups

3 Using the bottom-up approach to molecular manufacturing has advantages over the top-down approach because:

A it produces better molecules.

B it requires less energy.

C it is on an atomic level.

D it produces more wastage.

4 Which of the following statements regarding nanoparticles is correct?

A Nanoparticles have properties different from the same particles on a macro scale.

B Nanoparticles have not yet been used commercially.

C Nanoparticles are larger than a millimetre in diameter.

D Nanoparticles are smaller than atoms.

5 Explain the difference between the L- and D-forms of amino acid isomers.

___

___

___

___

6 A nanosensor is a biosensor that has been nanosized. A great deal of research is being carried out on the use of nanosensors to detect a wide range of diseases by the chemicals they produce.

A particularly promising field is the early warning system for people with diabetes. One of the symptoms of low blood sugar is the increased production of propanone.

The following information describes the cantilever array method of detecting propanone in the bloodstream.

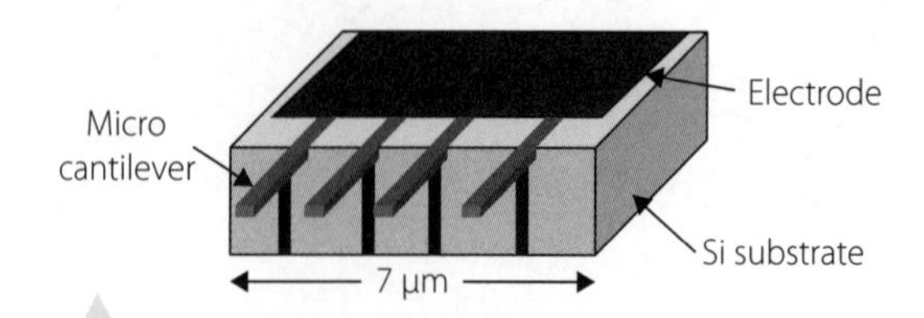

FIGURE 23.4.1 Cantilever array for the detection of propanone

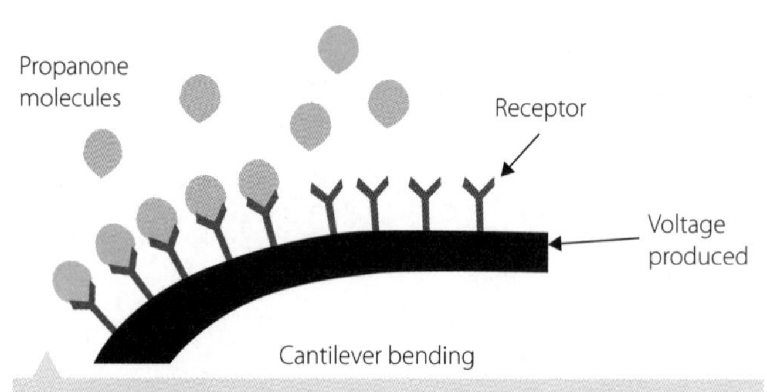

FIGURE 23.4.2 Bending of a cantilever due to bonding with propanone

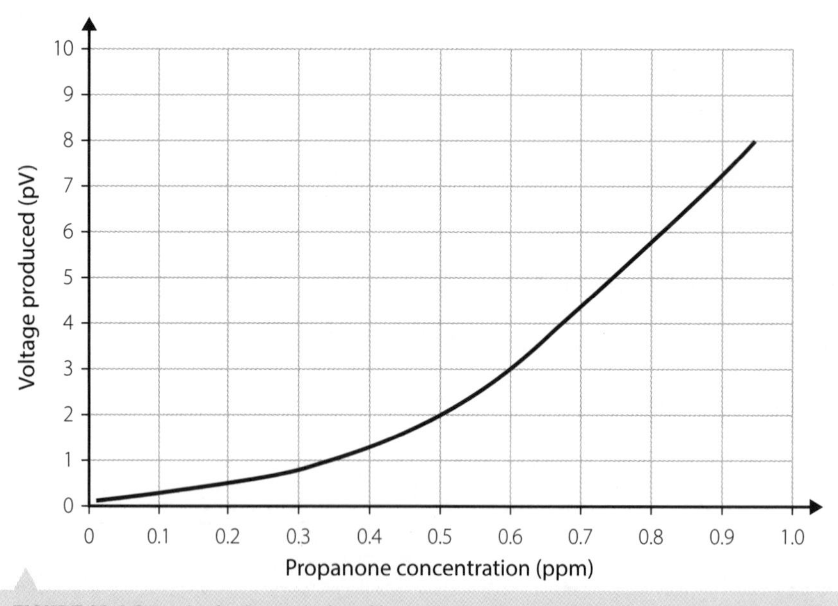

FIGURE 23.4.3 Graph of voltage produced by a cantilever array against propanone concentration

9780170412476

Write a description of how a cantilever array can detect propanone in a person with diabetes.

CHEMISTRY UNITS 3 AND 4

MULTIPLE CHOICE

1 A pure hydrocarbon is burned in oxygen. The products are carbon dioxide (1.32 g) and water (0.68 g). What is the empirical formula of this hydrocarbon?

A CH_4

B C_2H_5

C CH_2

D C_4H_5

2 In the production of sulfuric acid, an important step is the conversion of sulfur dioxide (SO_2) to sulfur trioxide (SO_3):

$$SO_2(g) + O_2(g) \rightleftharpoons 2SO_3(g)\ \Delta H = -196\ \text{kJ mol}^{-1}$$

Which of the following sets of conditions would give the best equilibrium yield of sulfur trioxide?

A 800°C and 5 atm pressure

B 25°C and 1 atm pressure

C 25°C and 5 atm pressure

D 800°C and 1 atm pressure

3 The acid dissociation constant of ethanoic acid is 1.7×10^{-5}, whereas that of hydrofluoric acid, HF, is 6.3×10^{-4}. When two solutions of these acids of equal concentration are compared, the solution of HF will have:

A a lower pH as it is more ionised.

B a lower pH as it is less ionised.

C a higher pH as it is more ionised.

D a higher pH as it is less ionised.

4 150 mL of 0.20 M hydrochloric acid, 300 mL of 0.10 M hydrochloric acid and 250 mL of 0.50 M hydrochloric acid are mixed. The concentration of hydrochloric acid in the resulting solution is:

A 0.264 M.

B 0.132 M.

C 0.528 M.

D 0.0528 M.

5 Which of the following is a redox reaction?

A $HNO_3(aq) + NaOH(aq) \rightarrow NaNO_3(aq) + H_2O(l)$

B $HNO_3(aq) + Na(s) \rightarrow NaNO_3(aq) + H_2(g)$

C $2HNO_3(aq) + Na_2CO_3(aq) \rightarrow 2NaNO_3(aq) + CO_2(g) + H_2O(l)$

D $2HNO_3(aq) + Na_2O(s) \rightarrow 2NaNO_3(aq) + H_2O(l)$

6 Four half-cells are set up as shown:

Half-cell I: An electrode of metal Z in a solution of Z^{2+} ions

Half-cell II: An electrode of metal X in a solution of X^{2+} ions

Half-cell III: An electrode of metal P in a solution of P^{+} ions

Half-cell IV: An electrode of metal Q in a solution of Q^{3+} ions

The half-cells are connected together in pairs, as shown, and the subsequent voltage generated is measured.

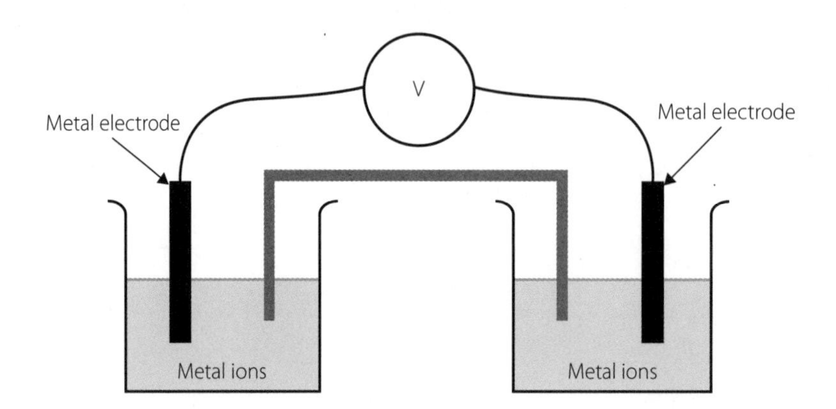

The results are shown below.

Half-cells used	Positive electrode	Negative electrode	Voltage (V)
I and III	P	Z	1.01
II and IV	X	Q	1.23
I and IV	P	Q	0.73
II and III	X	Z	1.53

Which of the following is the correct order of increasing reductant strength?

A Z, P, Q, X

B Z, Q, P, X

C X, P, Q, Z

D X, Q, P, Z

7 Which of the following statements applies to compounds in the same homologous series?

I They have the same molecular formula but different structural formulas.

II They contain the same functional group.

III They have similar chemical properties.

IV They have identical physical properties.

A II only

B I, III and IV

C II and III only

D II, III and IV only

8 An alkene has the molecular formula C_6H_{12}. When it is reacted with hydrogen and a nickel catalyst, 2-methyl pentane is formed. Which of the following could NOT be the structure of the alkene?

A $(CH_3)_2C{=}CHCH_2CH_3$

B $CH_3CH_2CH_2C(CH_3){=}CH_2$

C $CH_3CH_2CH_2CH{=}CHCH_3$

D $(CH_3)_2CHCH_2CH{=}CH_2$

9780170412476

9 1-pentene and 2-pentene are structural isomers and have the same molecular formula, C_5H_{10}. If both compounds are reacted with hydrogen chloride, HCl, in separate reactions, the products will:

A be structural isomers.

B be the same compound.

C differ in formula by $-CH_2-$ but remain in the same homologous series.

D be geometric isomers.

10 Consider the following types of interactions between atoms and molecules.

I Hydrogen bonds

II Dispersion forces

III Dipole–dipole interactions

IV Covalent bonds,

The tertiary structures of proteins are held together by:

A II and III.

B I, II and III.

C I and IV.

D all of the above.

11 A polyunsaturated fat is hydrolysed, forming glycerol and a polyunsaturated fatty acid. Which of the following could be the structural formula of the fatty acid formed?

A $C_{19}H_{39}COOH$

B $C_{16}H_{31}COOH$

C $C_{13}H_{21}COOH$

D $C_{14}H_{27}COOH$

12 Starch mainly consists of amylose and amylopectin forms, both of which are polymers of glucose, $C_6H_{12}O_6$. A particular form of amylopectin has a molar mass of 2.680×10^5 g mol^{-1}. A molecule of this amylopectin can be described as:

A an addition polymer of 1654 glucose molecules.

B an addition polymer of 2016 glucose molecules.

C a condensation polymer of 2016 glucose molecules.

D a condensation polymer of 1654 glucose molecules.

13 In the mass spectrum of dipropyl ether, $CH_3CH_2OCH_2CH_3$, significant peaks are found with *m/z* ratio values of 15, 29, 45 and 74. The peak with an *m/z* ratio of 29 is due to the existence of:

A CH_3CH_2.

B $CH_3CH_2^+$.

C CH_3CH_2O.

D $CH_3CH_2O^+$.

14 25 mL of 8 M HNO_3 is added to 50 mL of deionised water. The concentration of the diluted acid is:

A 2.67 M.

B 4 M.

C 3.00 M.

D 3.67 M.

15 The reaction below is allowed to reach equilibrium at 600°C.

$$2CO(g) + 5H_2(g) \rightleftharpoons 2H_2O(g) + C_2H_6(g)\ \Delta H = +256\ kJ\ mol^{-1}$$

The temperature is raised and subsequently the amount of H_2O changes by 0.40 mol. Which of the following best describes the changes that occurred?

	H_2O	C_2H_6	CO	H_2
A	Decrease by 0.40 mol	Decrease by 0.20 mol	Increase by 0.40 mol	Increase by 1 mol
B	Increase by 0.40 mol	Increase by 0.20 mol	Decrease by 0.40 mol	Decrease by 1 mol
C	Decrease by 0.40 mol	Decrease by 0.40 mol	Increase my 0.40 mol	Increase by 0.40 mol
D	Increase by 0.40 mol	Decrease by 0.20 mol	Decrease by 0.20 mol	Increase by 1 mol

16 Calcium hydroxide is a strong alkali. The pH at 25°C of a 0.0060 M solution of calcium hydroxide is:

A 1.92.

B 2.00.

C 12.08.

D 12.00.

17 Which one of the following is true for both galvanic cells and electrolytic cells?

A The cathode is negative in both cells.

B Oxidation occurs at the positive electrode in both cells.

C Anions migrate to the cathode in both cells.

D Reduction occurs at the cathode in both cells.

18 The compound 3-hydroxypropanoic acid ($HO(CH_2)_2COOH$) reacts to form a polyester. The polymer structure is best represented by:

A $—O(CH_2)_2CO—O(CH_2)_2CO—O(CH_2)_2CO—O(CH_2)_2CO—O(CH_2)_2CO—$.

B $—O(CH_2)_2O—OC(CH_2)_2CO—O(CH_2)_2O—OC(CH_2)_2CO—O(CH_2)_2O—$.

C $—O—(CH_2)_2—O—(CH_2)_2—O—(CH_2)_2—O—(CH_2)_2—O—(CH_2)_2—O—(CH_2)_2—O—(CH_2)—$.

D $—CO—(CH_2)_2—CO—(CH_2)_2—CO—(CH_2)_2—CO—(CH_2)_2—CO—(CH_2)_2—CO—(CH_2)_2—$.

19 The IR spectrum of lactic acid, 2-hydroxypropanoic acid, shows a sharp, strong absorption at approximately 1750 cm^{-1}. This is caused by the:

A O—H bond.

B C—C bond.

C C=O bond.

D C—H bond.

20 Which of the following best describes geometric isomers?

A Same molecular formula, different structural arrangement of atoms

B Same empirical formula, different molecular formula

C Same structural formula, different arrangement of atoms in space

D Same molecular formula, different molecular shape

9780170412476

SHORT ANSWER AND COMBINATION-RESPONSE QUESTIONS

1 Reactions I and II can be used to produce chlorine.

Reaction I: $PCl_5(g) \rightleftharpoons PCl_3(g) + Cl_2(g)$ $\Delta H = +420$ kJ mol^{-1}

Reaction II: $2HCl(g) \rightleftharpoons H_2(g) + Cl_2(g)$ $\Delta H = +322$ kJ mol^{-1}

a In the following table, identify the expected effect of each change on the equilibrium yield of chlorine (Increase, Decrease or No change).

Change to Reaction I and Reaction II	Effect of the change on the chlorine yield in Reaction I	Effect of the change on the chlorine yield in Reaction II
Increase in volume of the reaction vessel		
Reduction in temperature		
Increase in concentration of PCl_3 (Reaction I) or H_2 (Reaction II)		

b Using Le Chatelier's principle, explain the effect of increasing the volume (at constant temperature) of the reaction vessel on the equilibrium yield of chlorine in each reaction.

c Explain the effect of reducing the temperature on the equilibrium yield of chlorine in Reaction I with reference to the rates of the forward and reverse reactions.

d The equilibrium constant for Reaction I at 250°C is 1.05. Calculate the concentration of Cl_2 in mol L^{-1} that will result if 5.50 moles of PCl_5 is placed in a 20 L vessel at 250°C.

2 a Write an equation for the reaction of propanoic acid with water.

b Write an equilibrium expression for the acid dissociation constant K_a for the reaction in part **a**.

c A solution is prepared by adding 0.600 moles of propanoic acid, CH_3CH_2COOH, and 0.200 moles of potassium propanoate, CH_3CH_2COOK, to 3 L water. It can be assumed that the equilibrium concentrations of each reagent are equal to their initial concentrations. The acid dissociation constant for propanoic acid is 1.35×10^{-5} M.

i Calculate the pH of this solution.

9780170412476

ii How would the pH of this solution compare to that of a solution where no potassium propanoate had been added? Explain your reasoning.

______

______

______

______

3 A student must accurately determine the concentration of a solution of nitric acid by titration against a standard solution of sodium carbonate. The first step in the experiment is to dilute 100 mL of a 1 M stock solution of sodium carbonate. However, instead of using distilled water in the dilution, the student mistakenly adds 900 mL of 0.01110 M hydrochloric acid, HCl, solution.

a Write an equation to show the reaction that occurs in the 1 L volumetric flask.

______

______

b Calculate the concentration of the sodium carbonate that remains after the student added the HCl, to the appropriate number of significant figures.

______

______

______

______

______

______

______

______

______

______

The student then uses this contaminated solution of sodium carbonate to determine the accurate concentration of the unknown nitric acid solution.

c Will the calculated concentration of the nitric acid solution be greater or smaller than the true value? Justify your answer.

______

______

______

______

______

______

4 Sulfur dioxide is an airborne pollutant that can cause acid rain. The concentration of SO_2 in a sample of air can be determined by a back titration with potassium permanganate, $KMnO_4$, in the following reaction:

$$2MnO_4^-(aq) + 5SO_2(g) + 2H_2O(l) \rightarrow 2Mn^{2+}(aq) + 5SO_4^{2-}(aq) + 4H^+(aq)$$

The amount of excess permanganate remaining after reaction with the SO_2 is then determined by titration with a standard solution of Fe^{2+} as determined by the following equation:

$$MnO_4^-(aq) + 5Fe^{2+}(aq) + 8H^+(aq) \rightarrow Mn^{2+}(aq) + 5Fe^{3+}(aq) + 4H_2O(l)$$

In an experiment, 5 m^3 of polluted air was passed through 100 mL of 0.03020 M $KMnO_4$. The $KMnO_4$ that did not react with the SO_2 was reduced by reaction with 25.85 mL of 0.1480 M Fe^{2+} solution.

a Calculate the number of moles of excess MnO_4^- remaining after reaction with the SO_2.

____

____

____

____

b Calculate the number of moles of MnO_4^- that reacted with the SO_2 in the polluted air.

____

____

____

____

____

____

c Use this information to calculate the concentration of SO_2 in the polluted air in g m^{-3}.

____

____

____

____

____

____

5 A galvanic cell is constructed as shown in the diagram below:

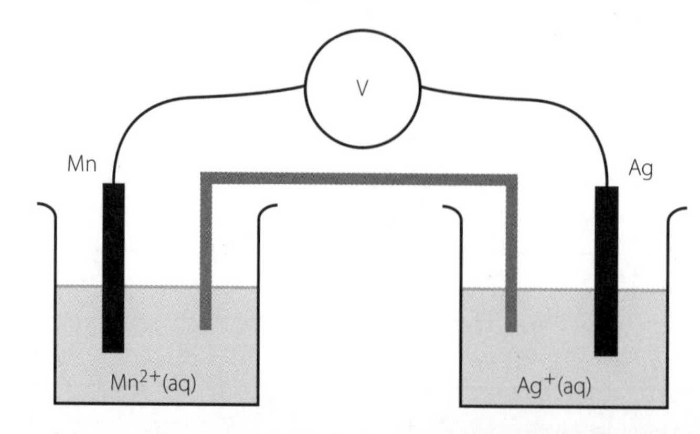

9780170412476

a Use the electrochemical series to add the following labels to the diagram above.

i The direction of electron flow in the external circuit

ii The direction of positive ion flow in the salt bridge

iii The polarity of each electrode

iv The anode and cathode

b Write half-equations and an overall ionic equation that is taking place in this galvanic cell.

Oxidation: ____________________

Reduction: ____________________

Overall equation: ____________________

c List two factors that need to be taken into account when selecting an appropriate substance to use in the salt bridge.

____

____

6 Three electrolytic cells are connected as shown and a steady current is passed through the system. Aqueous solutions of the substances shown are used as the electrolytes.

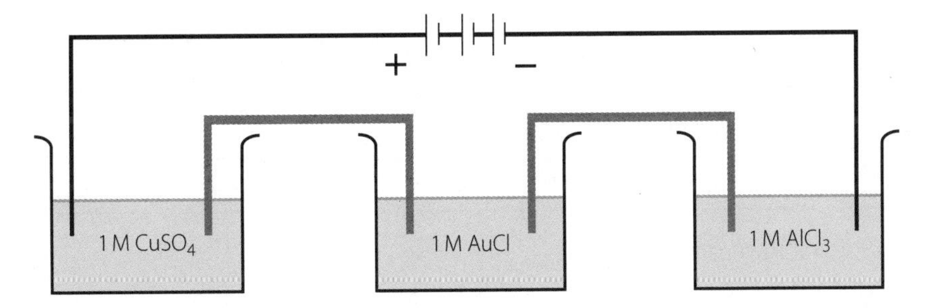

a Write the equations for the reactions occurring at the cathode in each of the three cells.

____

____

____

b If 0.12 moles of copper is deposited at the cathode of the first cell, how much material would you expect to be deposited in the gold and aluminium cells?

____

____

____

____

7 The flow chart below shows a reaction pathway used to convert an alkene A into an ester with the semi-structural formula of $(CH_3)_2CH_2CH_2CO_2CH_3$.

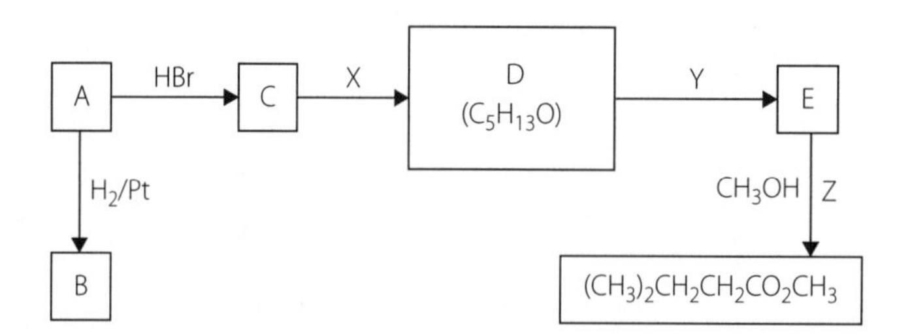

a Write the semi-structural formulas of the compounds **A** to **E**.

A ______________________________

B ______________________________

C ______________________________

D ______________________________

E ______________________________

b Write the formulas of the reagents labelled **X**, **Y** and **Z**.

X ______________________________

Y ______________________________

Z ______________________________

c Write the semi-structural formula of an isomer of **A**.

____

d Name the type of reaction used to convert:

i **A** to **C** ______________________________

ii **C** to **D** ______________________________

iii **D** to **E** ______________________________

8 The following compound is found in some plant oils:

$H_2C-O-C(=O)-C_{18}H_{29}$

$HC-O-C(=O)-C_{17}H_{31}$

$H_2C-O-C(=O)-C_{15}H_{21}$

The compound has the molecular formula $C_{56}H_{86}O_6$ and has a molecular mass of 854.

a How many C=C double bonds are there in the molecule? Explain your answer.

____

____

The iodine number is often used to indicate the amount of unsaturation in a molecule. It is the number of grams of iodine that reacts with 100 g of the molecule in an addition reaction. One molecule of iodine will add to each C=C bond.

9780170412476

b Calculate the iodine number for the oil in part **a**.

c Write the semi-structural formulas of the products formed when the oil is hydrolysed.

EXTENDED RESPONSE QUESTION

Nylon is a synthetic polyamide polymer with a wide range of uses in society.

$$-[-CO-CH_2-CH_2-CH_2-CH_2-CO-NH-CH_2-CH_2-CH_2-CH_2-CH_2-CH_2-NH-]_n-$$

Nylon is a very strong, unreactive substance that is used for making strong fibres which can be used for clothing and rope making.

Compare and contrast nylon with a polypeptide polymer that is naturally occurring in proteins. Include in your discussion:

- a comparison of the reactions by which each is formed and the reactants used
- a comparison of the tertiary structures formed by nylon and a protein and explain the reasons for the differences
- a comparison of the likely biodegradability of nylon compared to that of naturally occurring proteins and an explanation of your reasoning.

ANSWERS

CHAPTER 1 CHEMICAL EQUILIBRIUM

1.1 IMPORTANT TERMS

1 Crossword answers

Across	Down
3 System	**1** Surroundings
4 Equilibrium	**2** Reversible
7 Chemical	**5** Activated
9 Dynamic	**6** Closed
10 Physical	**8** Enthalpy
12 Open	**11** Activation

2 Answers will vary.

1.2 ETHANOIC ACID AND METHANOL IN AQUEOUS SOLUTION

1 and **3** Graphs of ethanoic acid and methanol concentration

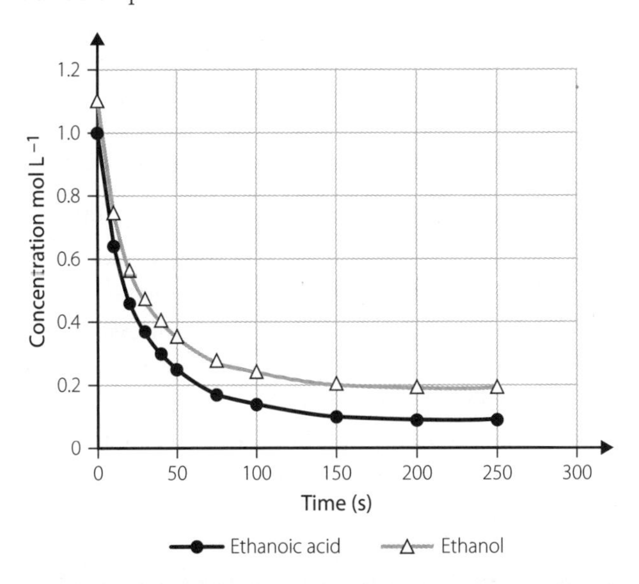

TIME $\frac{t}{(s)}$	$[CH_3COOH]$ (mol L^{-1})	CHANGE IN $[CH_3COOH]$ CONCENTRATION (mol L^{-1})	CHANGE IN $[CH_3OH]$ CONCENTRATION (mol L^{-1})
0	1	**0**	**1.1**
10	0.64	**0.36**	**0.74**
20	0.46	**0.54**	**0.56**
30	0.37	**0.63**	**0.47**
40	0.3	**0.7**	**0.4**
50	0.25	**0.75**	**0.35**
75	0.17	**0.83**	**0.27**
100	0.14	**0.86**	**0.24**
150	0.1	**0.9**	**0.2**
200	0.09	**0.91**	**0.19**
250	0.09	**0.91**	**0.19**

2 There is a greater range of change in the concentration early in the reaction; therefore, more data points are obtained to accurately plot the change.

4 Reaction reached equilibrium at approximately $t = 200$ s.

5 Yes, both forward and reverse reactions are occurring, but at the same rate, so there is no net change in concentration.

1.3 FORMING STALACTITES AND STALAGMITES

1 a As water drips over limestone in the presence of carbon dioxide, a forward reaction occurs and the calcium carbonate is eroded to form soluble calcium hydrogen carbonate. This drips through the rocks and dissolves the limestone.

b When the concentration of CO_2 in the air is less, the reverse reaction is favoured and so solid $CaCO_3$ is reformed, releasing CO_2 back into the air. The formation of stalactites and stalagmites occurs where the calcium hydrogen carbonate-rich water has dripped from a height to the ground.

2 Equilibrium is rarely reached as the concentration of carbon dioxide is changing due to the external conditions, favouring either the forward reaction if the CO_2 concentration is high, or the reverse reaction if the CO_2 concentration is low.

EVALUATION

1 A dynamic equilibrium exists for a reversible reaction when the forward and reverse reaction are occurring simultaneously but at the same rate, so that there is no net change in the concentrations of reactants and products.

2 Denotes a reversible reaction

3 The reaction ratio is 3:1 for hydrogen : nitrogen. Hydrogen is used up three times as quickly as nitrogen.

4 When the graphs are horizontal, the reaction has not stopped, a dynamic equilibrium has been reached where the forward and reverse reactions are occurring at the same rate. Therefore, there is no net change in the concentrations of reactants and products.

CHAPTER 2 FACTORS THAT AFFECT EQUILIBRIUM

2.1 IMPORTANT TERMS

1 Crossword answers

Across	Down
3 Chatelier	**1** Catalyst
5 Activation	**2** Collision
6 Enthalpy	**4** Endothermic
7 Concentration	
8 Exothermic	

2 Answers will vary.

2.2 CONCENTRATIONS OF SPECIES

1 Initially SCN^- and Fe^{3+} were mixed and SCN^- was at the higher concentration.

2 Concentrations of SCN^- and Fe^{3+} decrease initially, with that of $FeSCN^{2+}$ increasing. Although the starting concentrations vary, the rate of change of each concentration (represented by the gradient of the graphs) is the same for each reagent, and decreases to zero as equilibrium is reached.

3 An additional amount of SCN^- was added to the system. This is indicated by the sharp change in SCN^- concentration.

4 Le Chatelier's principle states that when a change is made to a system at equilibrium, the system adjusts to partially oppose the change. Therefore, as $[SCN^-]$ increases, the forward reaction is favoured over the reverse reaction as the system acts to reduce $[SCN^-]$. As Fe^{3+} is a reactant, and $FeSCN^{2+}$ a product of the forward reaction, $[Fe^{3+}]$ falls and $[FeSCN^{2+}]$ increases. Equilibrium is re-established when the rate of the reverse reaction rises until it is equal with that of the forward reaction.

2.3 BOILING WATER

1 The reaction being discussed is:

$H_2O(l) \rightleftharpoons H_2O(g)$

Boiling water in a pot on a stove will not come to equilibrium with the water vapour because the water vapour is dispersed throughout the room and therefore not present to condense and reform the liquid.

If a lid is added to the pot, the concentration of steam will build up to the point where the rate of the reverse condensation reaction increases such that it is equal with that of the forward reaction and an equilibrium is established.

2.4 EXPLAIN USING LE CHATELIER'S PRINCIPLE

1 Addition of sodium hydroxide solution, an alkali, causes [HOBr] to decrease in a neutralisation reaction. In order to minimise this change, the equilibrium shifts to the right, using more Br_2 in an attempt to replace the lost HOBr. Therefore, $[Br_2]$ decreases and the brown colour disappears. If HCl is added, then $[H^+]$ increases. In an attempt to minimise the increase in $[H^+]$ the equilibrium shifts to the left. This causes more Br_2 to be formed and therefore the brown colour to reappear.

2.5 CHANGING THE CONDITIONS

1 a The amount of SO_3 increases when partial pressure of oxygen is increased. An increase in reactant concentration causes the equilibrium to shift to the left in an attempt to minimise this change and therefore form more SO_3.

b There is a decrease in number of gas particles from right to left in this reaction. Therefore, if the volume of the system is increased, the concentration of gas particles decreases, and the system acts to oppose this change by shifting to the left. As a result, the amount of SO_3 decreases.

c The forward reaction is exothermic, as evidenced by the production of heat in this reaction. If the temperature is increased, the backward reaction, which is endothermic, will be favoured in order to absorb heat and reduce the temperature. Therefore, the amount of SO_3 decreases.

2.6 CHANGES TO PRESSURE IN GASEOUS REACTIONS

1 a Yes, the total pressure will have increased as the number of particles per unit volume will have increased.

b No, the concentrations of the reactant and product molecules have not changed, as the number of particles per unit volume is unchanged.

c Yes, the system is at equilibrium and is unchanged by the addition of nitrogen, an inert gas.

EVALUATION

1

	COLOUR AT NEW EQUILIBRIUM COMPARED TO INITIAL EQUILIBRIUM		$[Fe^{3+}]$ AT NEW EQUILIBRIUM COMPARED WITH INITIAL EQUILIBRIUM	
CHANGES MADE	MORE RED	LESS RED	INCREASED	DECREASED
HPO_4^{2-} is added which forms colourless $FeHPO_4^+$		✓		✓
Addition of a large volume of water		✓		✓
$Fe^{3+}(aq)$ is added	✓		✓	
Hg^{2+} ions are added which forms a precipitate of $Hg(SCN)_2$		✓	✓	

2

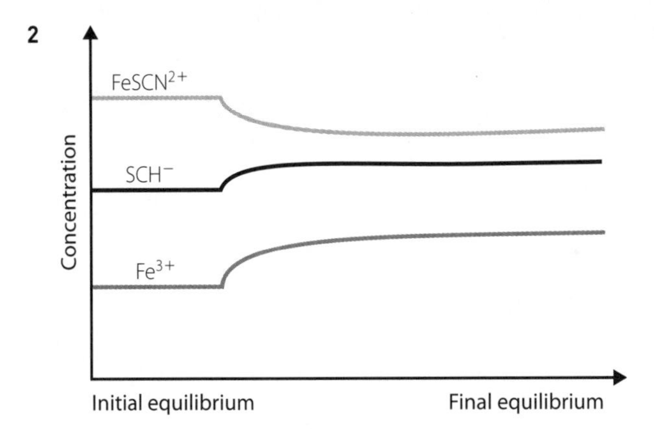

CHAPTER 3 EQUILIBRIUM CONSTANTS

3.1 IMPORTANT TERMS

1 Crossword answers

Across	Down
1 Constant	2 Temperature
3 Homogeneous	4 Expression
5 Quotient	
6 Heterogeneous	
7 Left	

2 Answers will vary.

3.2 EQUILIBRIUM CONSTANT EXPRESSIONS

1 $K - \frac{[CH_3COO^-][H_3O^+]}{[CH_3COOH][H_2O]}$

2 $K = \frac{[Mg^{2+}][OH^-]^2}{[Mg(OH)_2]}$

3 $K = \frac{[SO_3]^2}{[SO_2]^2[O_2]}$

4 $K = \frac{[Fe_2O_3]^2}{[Fe]^4[O_2]^3}$

3.3 AT EQUILIBRIUM OR NOT?

1 Data from the table shows the equilibrium constant as being:

$$K = \frac{0.56 \times 0.56}{0.44 \times 0.44}$$

$$= 1.620$$

In the second vessel, the concentration of $[H_2O(g)]$ has decreased:

$$0.600 - 0.456$$

$$= 0.144$$

Therefore, the [CO] will also have decreased by 0.144, and $[H_2]$ and $[CO_2]$ will have increased by 0.144.

The table can be updated to show:

	INITIAL CONCENTRATION (M)	CHANGE IN CONCENTRATION (M)	CONCENTRATION AFTER TIME HAD ELAPSED (M)
$H_2(g)$	0.200	+0.144	0.344
$CO_2(g)$	0.400	+0.144	0.544
$H_2O(g)$	0.600	−0.144	0.456
CO(g)	0.800	−0.144	0.656

Therefore, the reaction quotient is now:

$$K = \frac{0.456 \times 0.656}{0.344 \times 0.544}$$

$$= 1.598$$

The value of the reaction quotient is very close to the equilibrium constant, which means that the system is very close to, or at equilibrium.

3.4 CALCULATING K_c

1 and **2** There are many ways of solving these questions. For example,

SUBSTANCE	INITIAL NUMBER OF MOLES	CHANGE IN NUMBER OF MOLES	NUMBER OF MOLES AT EQUILIBRIUM	CONCENTRATION AT EQUILIBRIUM
H_2	1	$-\frac{3}{2} \times 0.078$	0.883	$\frac{0.883}{5} = 0.1766$
N_2	0.40	$-\frac{1}{2} \times 0.078$	0.361	$\frac{0.361}{5} = 0.0722$
NH_3	0	+0.078	0.078	$\frac{0.078}{5} = 0.0156$

3 $K = \frac{[H_2]^3[N_2]}{[NH_3]^2}$

$= \frac{0.1766^3 \times 0.0722}{0.0156^2}$

$= 1.634 \text{ M}$

3.5 USING EQUILIBRIUM CONSTANTS TO PREDICT QUANTITIES

1 At equilibrium,

$$\frac{[HI]^2}{[H_2][I_2]} = 50$$

Also, $[H_2] = [I_2]$

$= \frac{0.03}{0.5}$

$= 0.06$

Therefore, $[HI]^2 = 50 \times 0.06^2$

$= 0.18$

$[HI] = 0.424 \text{M}$

The number of moles of HI present at equilibrium will be $0.424 \times 0.5 = 0.212$ moles.

EVALUATION

1

MOLECULE	MOLES AT START	CHANGE	MOLES AT EQUILIBRIUM	CONCENTRATION AT EQUILIBRIUM
NO_2	1.42	−0.6	0.82	0.41
N_2O_4	0	+0.3	0.3	0.15

$K = \frac{0.15}{0.41^2}$

$= 0.892 \text{ M}^{-1}$

2 Exothermic. The temperature has decreased and the equilibrium constant has increased, meaning that the forward reaction has been favoured.

9780170412476

CHAPTER 4 PROPERTIES OF ACIDS AND BASES

4.1 CONCEPT MAP

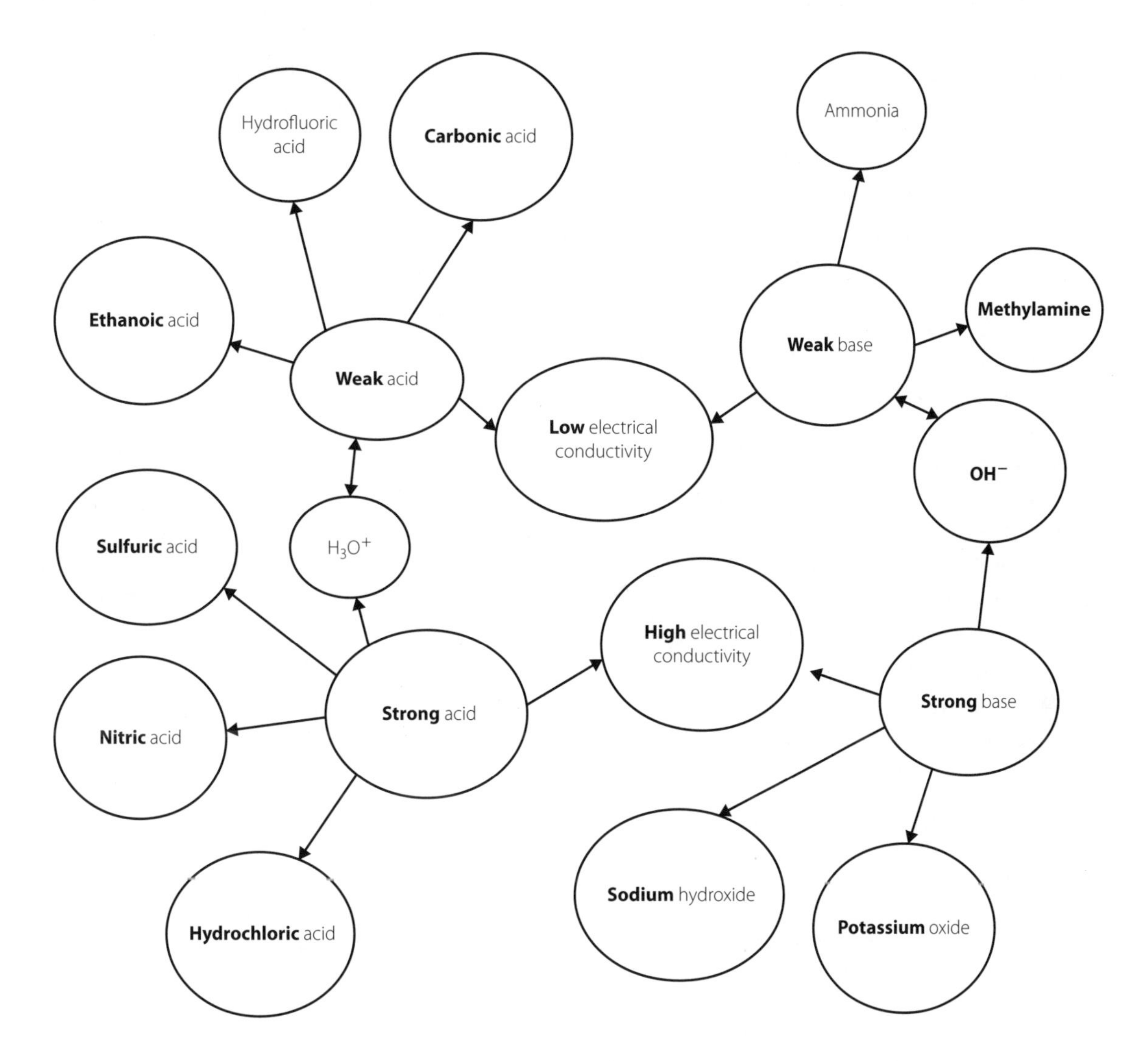

4.2 ACIDS AND BASES

1 a A substance that produces H^+ ions in solution

b A substance that produces OH^- ions in solution

c The ion that forms when a proton is donated to a water molecule (H_3O^+)

d A substance that can donate one proton to a base

e A substance that can donate two protons to a base

f A substance that can donate three protons to a base

g An acid in which only a small proportion of the molecules will donate a proton

h A base in which only a small proportion of the molecules dissociate in water

2 a A strong **base** is one that dissociates completely to produce OH^- ions in solution. Strong bases include **sodium hydroxide** and **potassium oxide**.

b A weak acid such as **ethanoic acid** only **partially** dissociates to produce **OH^-** in solution.

c A diprotic acid dissociates to produce **two hydronium** ions per molecule of acid.

4.3 ACID AND BASE STRENGTH

1 a **Sulfuric acid** dissociates completely to produce **hydronium** ions and sulfate ions.

b **Ethanoic acid** partially dissociates to produce **hydronium** ions and ethanoate ions.

c Phosphoric acid is a **triprotic** acid because it dissociates in **three** stages.

d Carbonic acid dissociates in two stages to give hydronium ions and **hydrogen carbonate** ions in the first stage, and hydronium ions and **carbonate** ions in the second stage.

4.4 CONCENTRATION VERSUS STRENGTH

1 a The **strength** of an acid or a base refers to the degree of dissociation that occurs and, therefore, the number of hydronium or hydroxide ions produced in solution.

b The **concentration** of an acid or a base describes how many molecules of an acid or base are present per unit volume of solution.

c Acids and bases can be strong or **weak**.

d Acids and bases can be concentrated or **dilute**.

EVALUATION

1 C **2** B **3** C **4** D

5 $H_3PO_4(aq) + H_2O(l) \rightarrow H_2PO_4^-(aq) + H_3O^+(l)$

dihydrogen phosphate ion

$H_2PO_4^-(aq) + H_2O(l) \rightarrow HPO_4^{2-}(aq) + H_3O^+(l)$

hydrogen phosphate ion

$HPO_4^{2-}(aq) + H_2O(l) \rightarrow PO_4^{3-}(aq) + H_3O^+(aq)$

phosphate ion

6 Nitric acid is a strong acid; all of its molecules dissociate to produce H_3O^+ ions in solution. Carbonic acid is a weak acid and even though there are far more carbonic acid molecules present in a 5 M solution, very few of those molecules will dissociate to produce H_3O^+ ions.

7 a 1 Only HCl is present. It is a strong acid and so it is fully dissociated. The large number of ions in solution gives a high conductivity.

2 As NaOH is added, neutralisation occurs between the OH^- ions and the H_3O^+ ions. Fewer ions in solution gives a lower conductivity.

3 All of the H_3O^+ ions have been neutralised. Very few ions in solution gives the lowest conductivity. This is the end point of the reaction.

4 As more NaOH is added the conductivity rises due to the presence of Na^+ and OH^- ions.

5 Maximum amount of NaOH has been added. The large number of Na^+ and OH^- ions gives a high conductivity.

b 2 M

End point occurred at 10 mL

$n_{HCl} = 1 \times 0.02$

$= 0.02\,M$

$= n_{NaOH}$

$c_{NaOH} = \frac{0.02}{0.01}$

$= 2\,M$

c (i) – (iii)

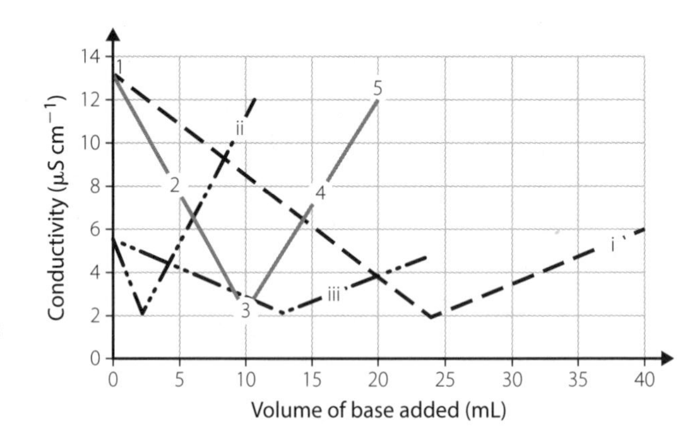

CHAPTER 5 pH SCALE

5.1 CONCEPT MAP

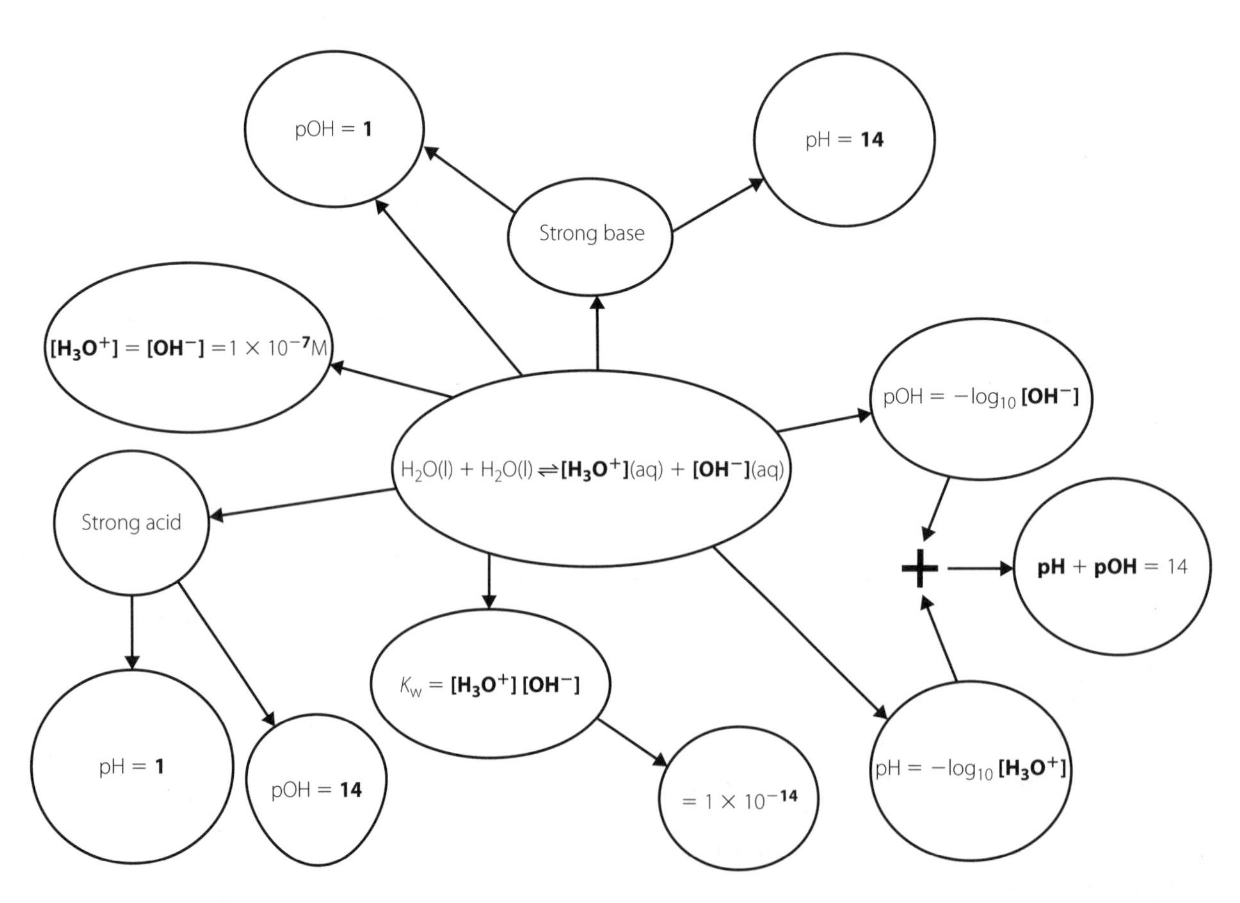

5.2 SELF-IONISATION OF WATER

1 a $[H_3O^+] = 0.74\,M$

$$[OH^-] = \frac{1 \times 10^{-14}}{0.74}$$
$$= 1.35 \times 10^{-14}\,M$$

b $[OH^-] = 1.39\,M$

$$[H_3O^+] = \frac{1 \times 10^{-14}}{1.39}$$
$$= 7.19 \times 10^{-15}\,M$$

c $[H_3O^+] = 2 \times 2.45 \times 10^{-4}$

$$= 4.9 \times 10^{-4}\,M$$
$$[OH^-] = \frac{1 \times 10^{-14}}{4.9 \times 10^{-4}}$$
$$= 2.04 \times 10^{-11}\,M$$

d $[OH^-] = 2 \times 1.15 \times 10^{-4}$

$$= 2.30 \times 10^{-4}\,M$$
$$[H_3O^+] = \frac{1 \times 10^{-14}}{2.30 \times 10^{-4}}$$
$$= 4.35 \times 10^{-11}\,M$$

5.3 pH SCALE

1 a $pH = -\log_{10} 0.0043$

$= 2.37$

$pOH = 14 - 2.37$

$= 11.63$

b $pOH = -\log_{10} (2 \times 0.98)$

$= -0.29$

$pH = 14 - (-0.29)$

$= 14.29$

c $pH = -\log_{10} (2 \times 1.76 \times 10^{-4})$

$= 3.45$

$pOH = 14 - 3.45$

$= 10.55$

d $pH = -\log_{10} 3.25$

$= -0.51$

$pOH = 14 - (-0.51)$

$= 14.51$

2 a $[H_3O^+] = 5.62 \times 10^{-12}\,M$

b $[H_3O^+] = 7.94 \times 10^{-14}\,M$

3 a $[OH^-] = 2.51 \times 10^{-11}\,M$

b $[OH^-] = 1.74 \times 10^{-14}\,M$

4

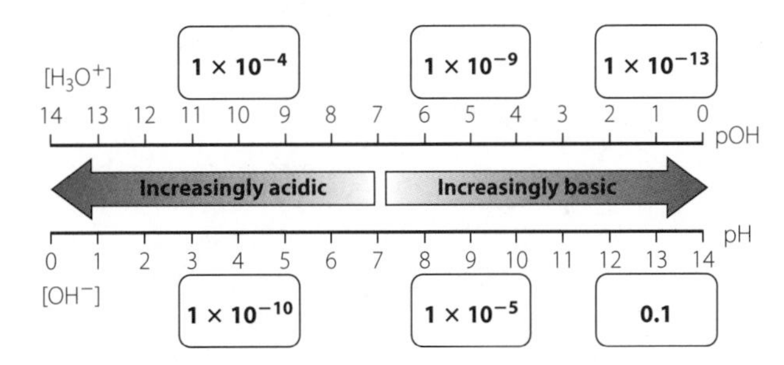

5 $pH = 13.26$

EVALUATION

1 A

2 C

3 B

4 D

5 1000 times more

6 pH 6.8

7 Highly concentrated acid and base solutions can have pH values that do not fall within the 1–14 range.

A 20 M solution of H_2SO_4 has a pH of $-\log_{10} (2 \times 20) = -1.6$

A 10 M solution of NaOH has a pH of $14 - (-\log_{10} 10) = 15$

8 13.0. To calculate this value, follow the steps in your student book, worked example 5.2.4.

9

$[H_3O^+]$	$[OH^-]$	pH	pOH	ACIDIC/BASIC
3.16×10^{-10}	**3.16×10^{-5}**	**9.5**	4.5	**Basic**
3.91×10^{-9}	**2.57×10^{-6}**	**8.41**	**5.59**	**Basic**
2.14×10^{-4}	4.7×10^{-11}	**3.67**	**10.33**	**Acidic**
0.16	**6.31×10^{-14}**	0.8	**13.2**	**Acidic**
1.99×10^{-13}	**0.05**	**12.7**	1.3	**Basic**
5.3×10^{-7}	**1.99×10^{-8}**	**6.3**	**7.7**	**Acidic**

CHAPTER 6 BRØNSTED–LOWRY MODEL

6.1 CONCEPT MAP

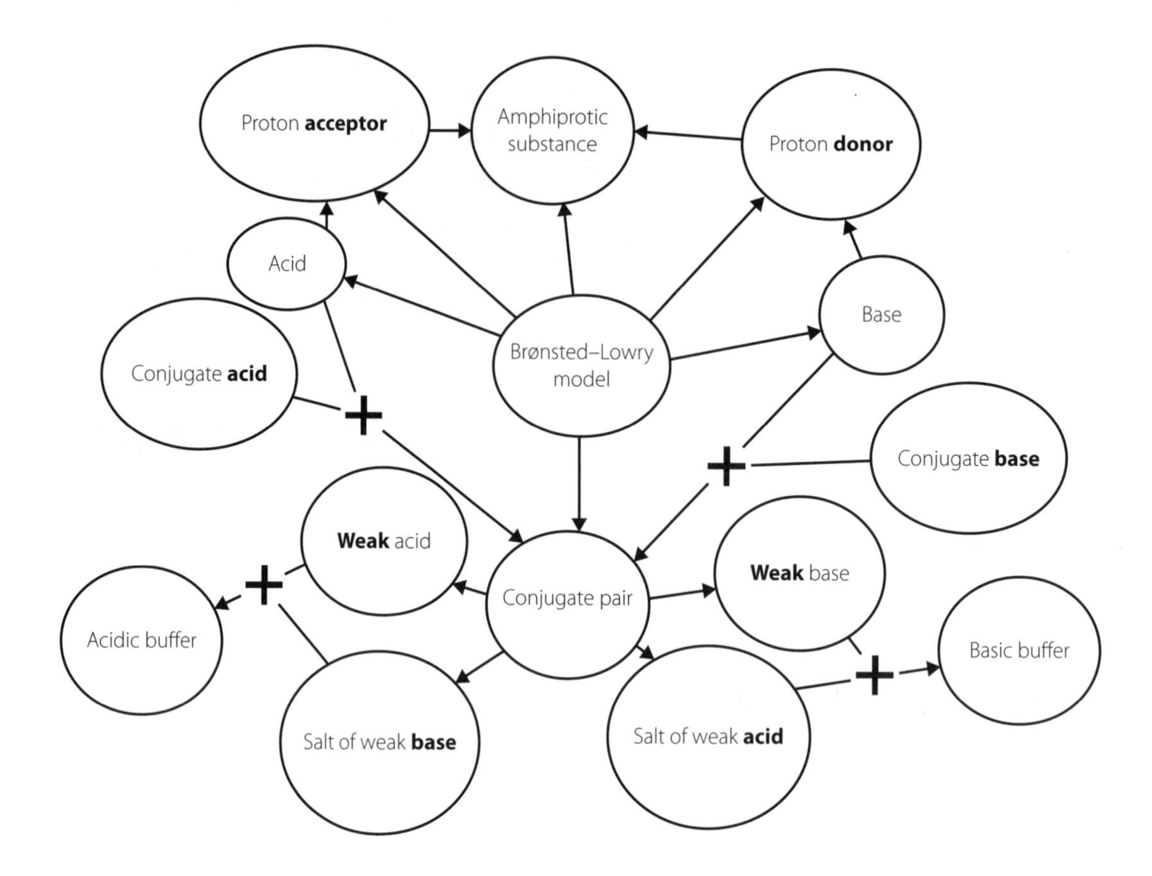

6.2 BRØNSTED–LOWRY MODEL

1 a

$HCl(aq) + NH_3(aq) \rightarrow NH_4^+(aq) + Cl^-(aq)$

A B CA CB

b

$H_2PO_4^-(aq) + HI(aq) \rightarrow NH_4^+(aq) + I^-(aq)$

A B CA CB

c

$H_2S(aq) + NH_2^-(aq) \rightarrow HS^-(aq) + NH_3(aq)$

A B CA CB

d

$CH_3COOH(aq) + CH_3CH_2NH_2(aq) \rightarrow CH_3COO^-(aq) + CH_3CH_2NH_3^+(aq)$

A B CA CB

6.3 AMPHIPROTISM

1 a Acid

$HCO_3^-(aq) + H_2O(aq) \rightarrow \mathbf{CO_3^{2-}(aq) + H_3O^+(aq)}$

Base

$HCO_3^-(aq) + H_2O(aq) \rightarrow \mathbf{H_2CO_3(aq) + OH^-(aq)}$

b Acid

$H_2PO_4^-(aq) + H_2O(aq) \rightarrow \mathbf{HPO_4^-(aq) + H_3O^+(aq)}$

Base

$H_2PO_4^-(aq) + H_2O(aq) \rightarrow \mathbf{H_3PO_4(aq) + OH^-(aq)}$

c Acid

$H_2NCH_2COOH(aq) + H_2O(aq) \rightarrow \mathbf{H_2NCH_2COO^-(aq) + H_3O^+(aq)}$

Base

$H_2NCH_2COOH(aq) + H_2O(aq) \rightarrow \mathbf{H_3N^+CH_2COOH(aq) + OH^-(aq)}$

6.4 BUFFERS

1 After an initial increase in pH due to the extra OH^- ions, the system shifts to the left as more NH_4^+ ions react, using up more OH^- ions and, therefore, restoring equilibrium.

EVALUATION

1 D **2** D **3** A **4** A

5 a

$NH_3(aq) + H_2O(l) \rightarrow NH_4^+(aq) + H_3O^+(aq)$

A B CB CA

9780170412476

b

$H_2SO_4(aq) + H_2O(aq) \rightarrow HSO_4^-(aq) + H_3O^+(aq)$
A B CB CA

$HSO_4^-(aq) + H_2O(l) \rightarrow SO_4^{2-}(aq) + H_3O^+(aq)$
A B CB CA

c

$HCO_3^-(aq) + H_2O(l) \rightarrow CO_3^{2-}(aq) + H_3O^+(aq)$
A B CB CA

d

$H_3BO_3(aq) + H_2O(l) \rightarrow H_2BO_3^-(aq) + H_3O^+(aq)$
A B CB CA

$H_2BO_3^-(aq) + H_2O(l) \rightarrow HBO_3^{2-}(aq) + H_3O^+(aq)$
A B CB CA

$HBO_3^{2-}(aq) + H_2O(l) \rightarrow BO_3^{3-}(aq) + H_3O^+(aq)$
A B CB CA

6 a A small amount of acid was added. The decrease in CH_3NH_2 concentration along with the increase in $CH_3NH_3^+$ and OH^- concentrations indicates that the system has shifted to the right. This occurred in response to the presence of H_3O^+. The production of extra OH^- ions neutralises the H_3O^+.

b A small amount of base was added. The increase in CH_3NH_2 concentration along with the decrease in $CH_3NH_3^+$ and OH^- concentrations indicates that the system has shifted to the left. This occurred in response to the presence of extra OH^- ions. By shifting to the left, the system uses up the extra OH^- ions.

c

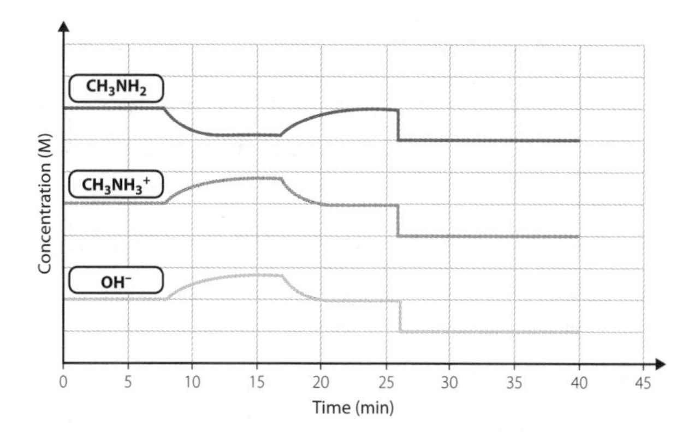

d The addition of water decreases the concentration of all species present in the system. However, the relative concentration of each species relative to one another remains the same.

CHAPTER 7 DISSOCIATION CONSTANTS

7.1 CONCEPT MAP

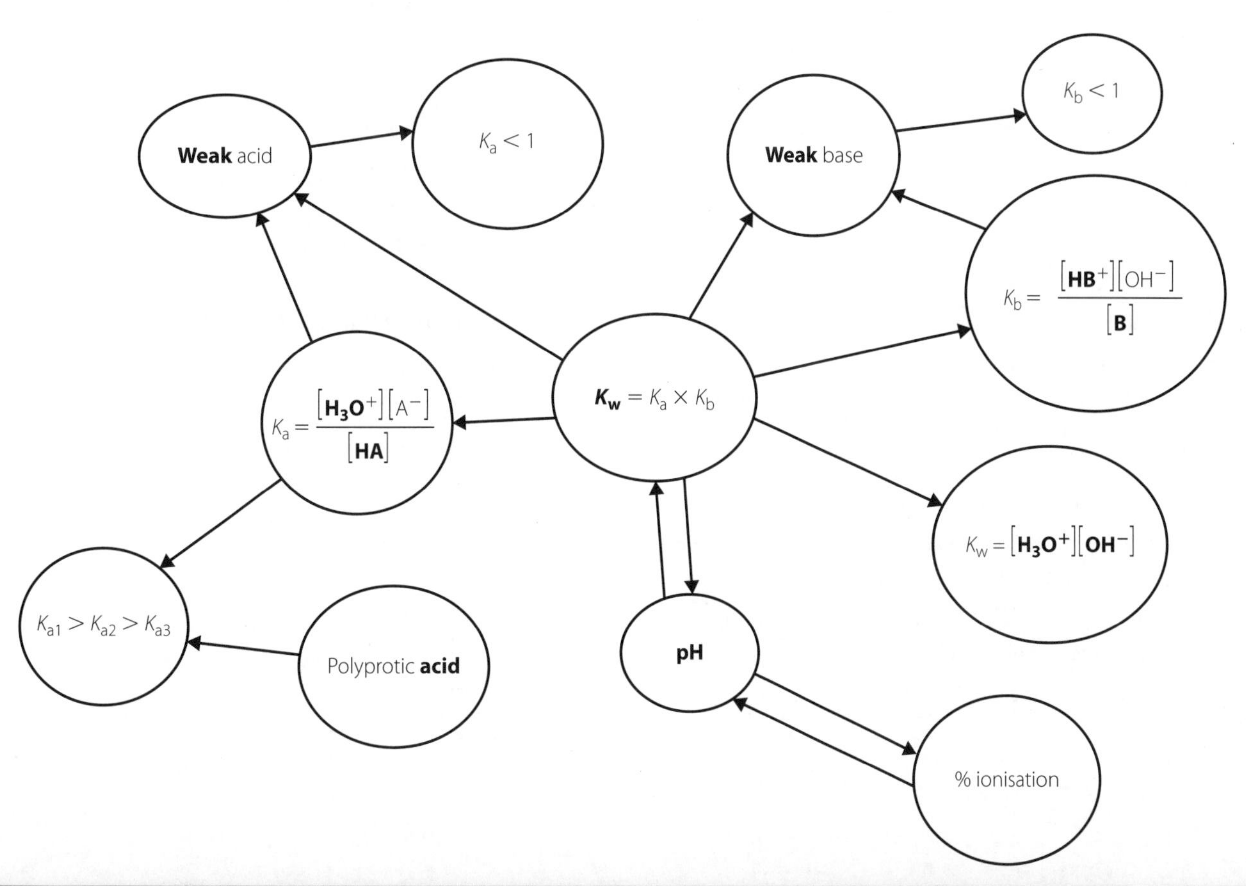

7.2 ACID DISSOCIATION CONSTANTS

1 When **weak** acids dissociate, an **equilibrium** is set up between the weak **acid** and its conjugate **base**. In this case, the equilibrium is shifted to the **left** because the amount of **conjugate base** is very small compared to the amount of **acid**. This gives rise to a very **small** K_a value.

2

Dissociation equation	Substance	Acid or base	Equilibrium expression
$CH_3COOH(aq) + H_2O(l) \rightleftharpoons CH_3COO^-(aq) + H_3O^+(aq)$	CH_3COOH	**Acid**	$K_a = \frac{[CH_3COO^-][H_3O^+]}{[CH_3COOH]}$
$CH_2ClCOOH(aq) + H_2O(l) \rightleftharpoons CH_2ClCOO^-(aq) + H_3O^+(aq)$	$CH_2ClCOOH$	**Acid**	$K_a = \frac{[CH_2ClCOO^-][H_3O^+]}{[CH_2ClCOOH]}$
$CH_3CH_2NH_2(aq) + H_2O(l) \rightleftharpoons CH_3CH_2NH_3^+(aq) + OH^-(aq)$	$CH_3CH_2NH_2$	**Base**	$K_b = \frac{[CH_3CH_2NH_3^+][OH^-]}{[CH_3CH_2NH_2]}$
$HBr(aq) + H_2O(l) \rightleftharpoons BR^-(aq) + H_3O^+(aq)$	HBr	**Acid**	$K_a = \frac{[Br^-][H_3O^+]}{[HBr]}$
$NH_3(aq) + H_2O(l) \rightleftharpoons NH_4^+(aq) + OH^-(aq)$	NH_3	**Base**	$K_b = \frac{[NH_4^+][OH^-]}{[NH_3]}$

7.3 ANALYSING EXPERIMENTAL DATA

1

SUBSTANCE	COLOUR IN UNIVERSAL INDICATOR	CONDUCTIVITY (μS)	IDENTIFICATION
P	Yellow	83	CH_3COOH
Q	Purple	728	$Ba(OH)_2$
R	Blue	62	CH_3NH_2
S	Red	665	H_2SO_4
T	Purple	413	NaOH
X	Orange	58	$H_2C_2O_4$
Y	Red	340	HCl
Z	Dark blue	94	$H_2NCH_2CH_2CH_2NH_2$

2 S, P, X and Y are acids.

S must be H_2SO_4 because it is red in universal indicator and, due to it being a strong diprotic acid, it puts more ions into a solution and thus has a higher conductivity.

Y must be HCl because it is red and, being a strong acid, has the next highest conductivity.

X must be $H_2C_2O_4$ because it is orange and, being diprotic, has a higher conductivity than P.

P must be CH_3COOH because it is yellow and has the lowest conductivity of the acids.

Q, R, T and Z are bases.

7.4 CALCULATIONS INVOLVING K_a AND K_b

1 $4.26 = -\log_{10}[H_3O^+]$

$[H_3O^+] = 5.49 \times 10^{-5}$ M

$$K_a = \frac{(5.49 \times 10^{-5})^2}{0.125}$$

$= 2.41 \times 10^{-8}$

2 $pOH = 14 - 9.84$

$= 4.16$

$[OH^-] = 6.92 \times 10^{-5}$ M

$$K_a = \frac{(6.92 \times 10^{-5})^2}{4.5^{-3}}$$

$= 1.06 \times 10^{-6}$

9780170412476

3 $1.34 \times 10^{-5} = \frac{[CH_3CH_2COO^-][H_3O^+]}{0.18}$

$2.4 \times 10^{-6} = [H_3O^+]^2$

$1.55 \times 10^{-3} = [H_3O^+]$

$pH = -\log_{10} 1.55 \times 10^{-3}$

$= 2.8$

4 $5.95 \times 10^{-4} = \frac{[CH_3CH_2NH_3^+][OH^-]}{0.05}$

$[OH^-] = 5.5 \times 10^{-3} M$

$pOH = 2.26$

$pH = 14 - 2.26$

$= 11.74$

5 $2.79 = -\log_{10} [H_3O^+]$

$1.62 \times 10^{-3} M = [H_3O^+] = [A^-]$

$\text{Percentage ionisation} = \frac{1.62 \times 10^{-3}}{0.89} \times 100\%$

$= 0.18\%$

6 $C_5H_5N(aq) + H_2O(l) \rightarrow C_5H_5NH^+(aq) + OH^-(aq)$

conjugate acid

$1.0 \times 10^{-14} = 1.5 \times 10^{-9} \times K_a$

$\frac{1.0 \times 10^{-14}}{1.5 \times 10^{-9}} = 6.67 \times 10^{-6}$

EVALUATION

1 A 2 C 3 B 4 B

5 At equilibrium, [HA] is the same as the initial concentration. This is reasonable because, being a weak acid, HA dissociates to a very small degree. This enables calculations to be much simpler.

$[H_3O^+]$ produced by the self-ionisation of water is negligible and has no effect on the calculations. Discounting H_3O^+ from the self-ionisation of water simplifies calculations.

6 Data on the acid–base nature of the substance is required. Conductivity readings simply give information on the number of ions in solution, not whether the solution is acidic or basic.

7 $C_6H_5COOH(aq) + H_2O(l) \rightarrow C_6H_5COO^-(aq) + H_3O^+(aq)$

$2.13 = -\log_{10} [H_3O^+]$

$[H_3O^+] = 7.41 \times 10^{-3} M$

$K_a = \frac{(7.41 \times 10^{-3})^2}{0.85}$

$= 6.46 \times 10^{-5}$

$K_b = \frac{1.0 \times 10^{-14}}{6.46 \times 10^{-5}}$

$= 1.55 \times 10^{-10}$

CHAPTER 8 ACID–BASE INDICATORS

8.1 ACID–BASE INDICATORS

1 In a basic solution, methyl orange is surrounded by OH^- ions. These neutralise the H_3O^+ ions. According to Le Chatelier's principle, the equilibrium will shift to the right in order to replace the lost H_3O^+ ions. This results in the formation of the yellow indicator.

8.2 THE RELATIONSHIP BETWEEN pH AND pK_a

1

ACID/BASE	K_a/K_b	PK_a/PK_b
Sulfurous acid	1.2×10^{-2}	**$pK_a = 1.90$**
Propanoic acid	**$K_a = 1.45 \times 10^{-5}$**	4.87
Ammonia	**$K_b = 1.78 \times 10^{-5}$**	4.75
Ethylamine	5.37×10^{-4}	**$pK_b = 3.27$**

2 $K_{ind} = \frac{[In^-][H_3O^+]}{[HIn]}$

3 $pK_{ind} = pK_a$

4

INDICATOR	K_a	pH RANGE
Methyl orange	2×10^{-4}	**2.7 – 4.7**
Phenol red	1×10^{-8}	**7 – 9**
Bromothymol blue	1×10^{-7}	**6 – 8**
Phenolphthalein	5×10^{-10}	**8.3 – 10.3**
Thymol blue	2×10^{-2}	**0.7 – 2.7**
Litmus	1×10^{-7}	**6 – 8**

8.3 INDICATOR COLOUR CHANGE

1 In the titration between hydrochloric acid and sodium hydroxide:

$HCl(aq) + NaOH(aq) \rightarrow NaCl(aq) + H2O(l)$

the **equivalence** point is when the exact amount of NaOH has been added to neutralise the HCl. The only substances present in the flask now are **NaCl** and **H_2O**. Technically, the conjugate **base** of the HCl is the **Cl^-** ion but this has no discernible **basic** properties because it is the conjugate **base** of a **strong** acid.

Similarly, the conjugate acid of the NaOH is the **Na^+** ion but this has no discernible **acid** properties because it is the conjugate **acid** of a **strong** base.

Therefore, the pH of the resultant solution is **7**. A suitable indicator for this titration would be **bromothymol blue** because **its useful pH range is between 6–8.**

2 $CH_3CH_2COOH(aq) + NaOH(aq) \rightarrow CH_3CH_2COO^-Na^+(aq) + H_2O(l)$

Species present at equivalence point:

$CH_3CH_2COO^-$: conjugate base of a weak acid. Therefore, it is a weak base, pH about 9 or 10.

Na^+: conjugate acid of a strong base. Therefore, it has no discernible properties.

H_2O: no discernible properties.

Phenolphthalein: useful pH range of 8.3–10.3 encompasses the predicted equivalence point of 9–10.

EVALUATION

1 B **2** D **3** C **4** D

5 Species present = Na^+, H_2O and CH_3COO^-. CH_3COO^- is the conjugate base of a weak acid, it has a pH of 9–10.

6 pH at equivalence point is about 11. This would be a strong base such as NaOH and a very weak acid such as HCN.

7 $HIn + H_2O \rightleftharpoons In^- + H_3O^+$

yellow blue

When acid is added, the system will shift to the left in order to remove the excess H_3O^+. A result of this is that the blue In^- form of the indicator changes to the yellow HIn form.

8 **a** For phenol red:

$HIn + H_2O \rightleftharpoons In^- + H_3O^+$

yellow red

Therefore, at pH 5: $[H_3O^+] = 1 \times 10^{-5}$ M

$$1 \times 10^{-8} = \frac{[In^-]\,1 \times 10^{-5}}{[HIn]}$$

$$1 \times 10^{-3} = \frac{[In^-]}{[HIn]}$$

[HIn] << [In^-] The dominant species is HIn. The solution is yellow.

b At pH 10.5: $[H_3O^+] = 3 \times 10^{-11}$ M

$$1 \times 10^{-8} = \frac{[In^-]\,3 \times 10^{-11}}{[HIn]}$$

$$333 = \frac{[In^-]}{[HIn]}$$

[HIn] << [In^-] The dominant species is In^-. The solution is red.

CHAPTER 9 VOLUMETRIC ANALYSIS

9.1 TYPES OF ACID–BASE TITRATION

1 At the **equivalence** point of a strong acid (HCl)/strong base (NaOH) titration, just enough NaOH has been added to neutralise the **HCl**. The solution remaining contains **Na^+** ions, **Cl^-** ions and **H_2O**. The pH of this solution is **7** because **Na^+ and Cl^- ions have no discernible acid–base properties**. An indicator with a useful pH range between **6** and **8** would be suitable for this type of titration.

2 At the **equivalence** point of a strong acid (HCl)/weak base (NH_3) titration, just enough NH_3 has been added to neutralise the **HCl**. The solution remaining contains **NH_4^+** ions and **Cl^-** ions. The pH of this solution is **5** because **the Cl^- ions have no acid–base properties, the H_2O is neutral, and the NH_4^+ ion is the conjugate acid of a weak base and so will be a weak acid**. An indicator with a useful pH range between **4** and **6** would be suitable for this type of titration.

3 At the **equivalence** point of a weak acid (CH_3COOH)/strong base (NaOH) titration, just enough NaOH has been added to neutralise the **CH_3COOH**. The solution remaining contains **Na^+** ions, **CH_3COO^-** ions and **H_2O**. The pH of this solution is **9** because **Na^+ ions have no discernible acid–base properties, H_2O is neutral and the CH_3COO^- ion is the conjugate base of a weak acid and so is a weak base**. An indicator with a useful pH range between **8** and **11** would be suitable for this type of titration.

9.2 ACID–BASE TITRATIONS

1

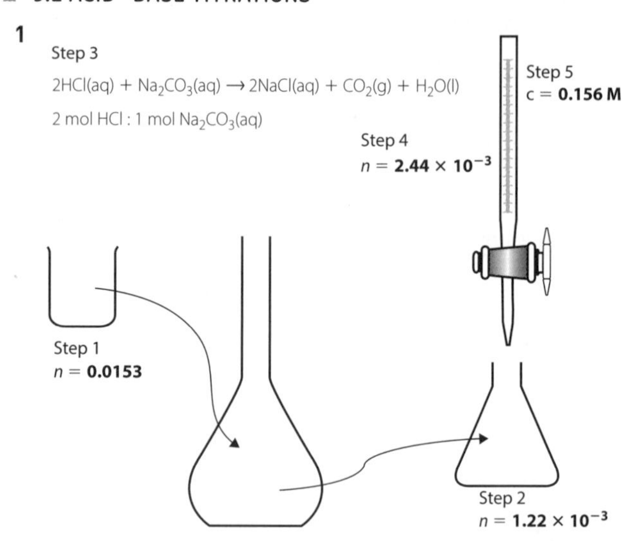

2 The substance to be analysed is insoluble. Impurities in the sample may interfere with a titration. The substance to be analysed is volatile and can evaporate slightly during titration. The end point is difficult to discern using a forward titration.

3 Ammonia is a volatile substance (it evaporates easily). During the time it takes for a conventional forward titration to take place, a small but significant amount of ammonia may have evaporated, creating an unacceptable error.

Worked example 2

a $0.103 \times \frac{20.00}{1000}$

$= 2.06 \times 10^{-3}$ mol (initial)

b $0.052 \times \frac{6.50}{1000}$

$= 3.38 \times 10^{-4}$ mol

c $n_{HCl} = 3.38 \times 10^{-4}$ mol (final)

d $2.06 \times 10^{-3} - 3.38 \times 10^{-4}$

$= 1.722 \times 10^{-3}$ mol (reacted)

e If there were 1.722×10^{-3} mol HCl (reacted) there must have been 1.722×10^{-3} mol NH_3 (present).

9780170412476

f Therefore, $n(NH_3)$ in 250 = 1.722×10^{-3} mol NH_3(present) × 12.5 = 0.0215 mol NH_3 in 250 mL

g $m(NH_3)$ = mol NH_3 in 250 mL × 17

$= 0.365\text{ g } NH_3$

h $\% NH_3 = \frac{0.365}{10.00} \times 100$

$= 3.65\%$

4 $n_{HCl\ initial} = 0.173 \times \frac{100.00}{1000}$

$= 0.0173\text{ mol}$

$n_{HCl\ final} = n_{NaOH}$

$= 0.114 \times \frac{21.60}{1000}$

$= 2.46 \times 10^{-3}\text{ mol}$

$n_{HCl\ reacted} = n_{HCl\ initial} - n_{HCl\ final}$

$= 0.0173 - 2.46 \times 10^{-3}$

$= 0.015\text{ mol}$

$n(Ca^{2+}) = \frac{1}{2} \times 0.015$

$= 7.5 \times 10^{-3}$

$m(Ca^{2+}) = 0.3\text{ g}$

$\% Ca^{2+} = \frac{0.3}{1.336} \times 100$

$= 22.45\%$

9.3 GRAPHS

1 and **2**

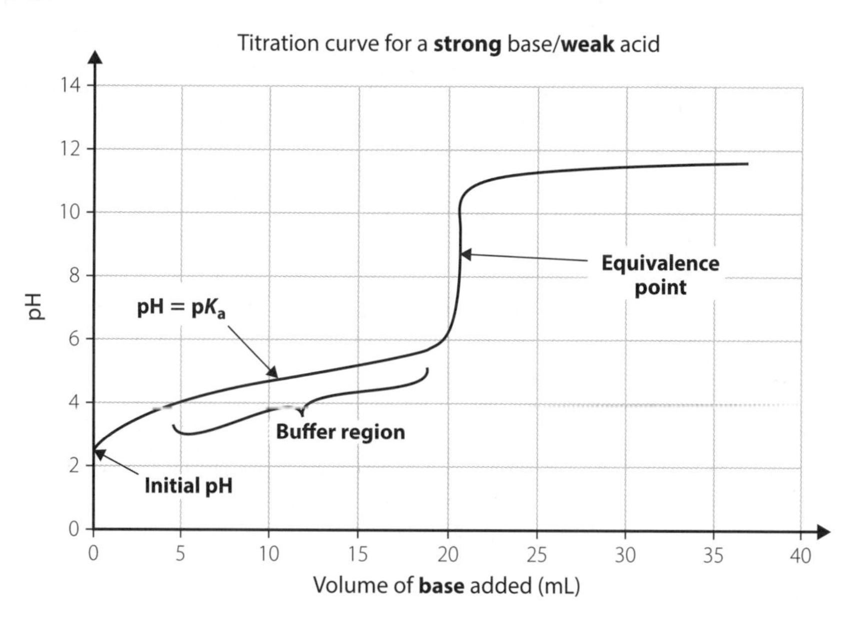

3 Methanoic acid/sodium hydroxide

Equivalence point = 8.5

$HCOOH(aq) + NaOH(aq) \rightarrow HCOO^-Na^+(aq) + H_2O(l)$

At equivalence point, the species present are Na^+ ions, $HCOO^-$ ions and water. Na^+ has no acid/base properties and H_2O is neutral. The methanoate ion, $HCOO^-$, is the conjugate base of a weak acid and so is a weak base with a pH of about 8.5.

EVALUATION

1 D **2** B **3** C **4** A

5 Magnesium hydroxide, $Mg(OH)_2$, commonly known as milk of magnesia, is only slightly soluble in water. This makes direct titration impossible. Analysis must be carried out by dissolving the $Mg(OH)_2$ in an excess of acid and then titrating the excess acid with a base such as NaOH.

6 Concordant titres are those produced during a titration that are within 0.1 mL of each other.

When performing calculations using titres, an average is used. The average can only be determined using concordant titres. Using titres that are outside of the 0.1 mL range introduces unacceptable error.

7 $n_{HCl(initial)} = \frac{0.317 \times 30.00}{1000}$

$= 9.51 \times 10^{-3}\text{ mol}$

$n_{HCl(final)} = n_{NaOH}$

$= \frac{0.04 \times 19.8}{1000}$

$= 7.92 \times 10^{-4}\text{ mol}$

$n_{HCl(reacted)} = 9.51 \times 10^{-3} - 7.92 \times 10^{-4}$

$= 8.72 \times 10^{-3}$

$n_{CaCO_3} = \frac{8.72 \times 10^{-3}}{2}$

$= 4.36 \times 10^{-3}\text{ mol}$

$m_{CaCO_3} = 4.36 \times 10^{-3} \times 100$

$= 0.436\text{ g}$

$\% CaCO_3 = \frac{0.436}{0.622} \times 100$

$= 70\%$

CHAPTER 10 REDOX REACTIONS

10.1 CONCEPT MAP

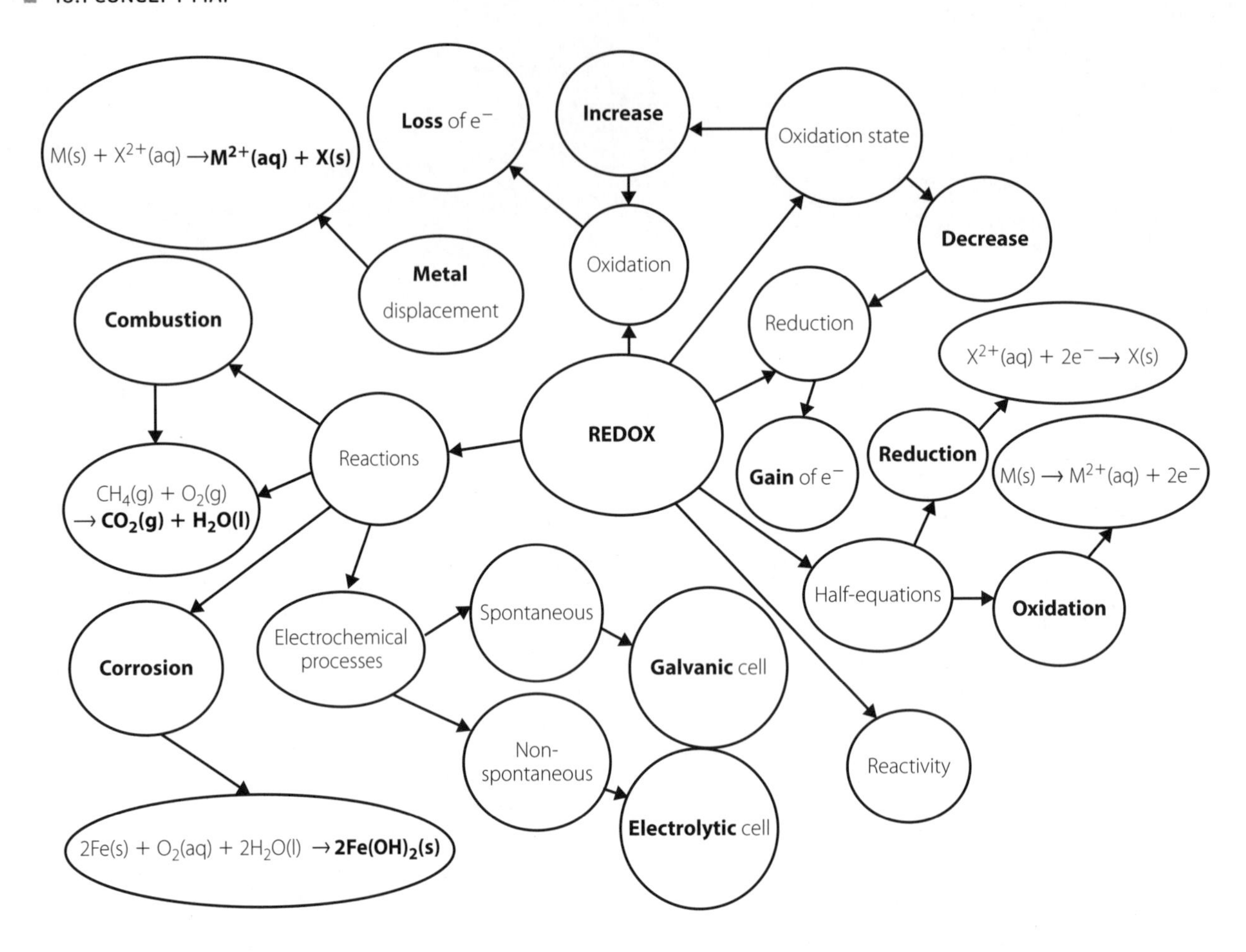

10.2 RANGE OF REDOX REACTIONS

1 $Q + T^{2+} \rightarrow Q^{2+} + T \therefore Q > T$

$Q \rightarrow Q^{2+} + 2e^-$ Oxidation

$T^{2+} + 2e^- \rightarrow T$ Reduction

$T + L^{2+} \rightarrow T^{2+} + L \therefore T > L$

$T \rightarrow L^{2+} + 2e^-$ Oxidation

$L^{2+} + 2e^- \rightarrow L$ Reduction

$Q + D^{2+} \rightarrow Q^{2+} + D \therefore Q > D$

$Q \rightarrow Q^{2+} + 2e^-$ Oxidation

$D^{2+} + 2e^- \rightarrow D$ Reduction

$L + D^{2+} \rightarrow L^{2+} + D \therefore L > D$

$L \rightarrow L^{2+} + 2e^-$ Oxidation

$D^{2+} + 2e^- \rightarrow D$ Reduction

$T + D^{2+} \rightarrow T^{2+} + D \therefore T > D$

$T \rightarrow T^{2+} + 2e^-$ Oxidation

$D^{2+} + 2e^- \rightarrow D$ Reduction

$Q + L^{2+} \rightarrow Q^{2+} + L \therefore Q > L$

$Q \rightarrow Q^{2+} + 2e^-$ Oxidation

$L^{2+} + 2e^- \rightarrow L$ Reduction

The strongest oxidising agent is the easiest to displace. D is the easiest to displace.

$D > L > T > Q$

2 $4Na(s) + O_2(g) \rightarrow 2Na_2O(s)$

$4Na(s) \rightarrow 4Na^+(s) + 4e^-$ Oxidation

$O_2(g) + 4e^- \rightarrow 2O^{2-}(s)$ Reduction

$2Ca(s) + O_2(g) \rightarrow 2CaO(s)$

$2Ca(s) \rightarrow 2Ca^{2+}(s) + 4e^-$ Oxidation

$O_2(g) + 4e^- \rightarrow 2O^{2-}(s)$ Reduction

$4Al(s) + 3O_2(g) \rightarrow 2Al_2O_3(s)$

$4Al(s) \rightarrow 4Al^{3+}(s) + 12e^-$ Oxidation

$3O_2(g) + 12e^- \rightarrow 6O^{2-}(s)$ Reduction

3 a Conditions for corrosion of iron to occur: Fe, O_2, H_2O. Refer to the 'corrosion' section of your student book, Section 10.1.

b Corrosion protection: Coat the iron bar with paint to prevent contact with oxygen and water.

Attach the iron bar to a more reactive metal, such as zinc, so that the more reactive metal corrodes instead of the iron bar.

9780170412476

c

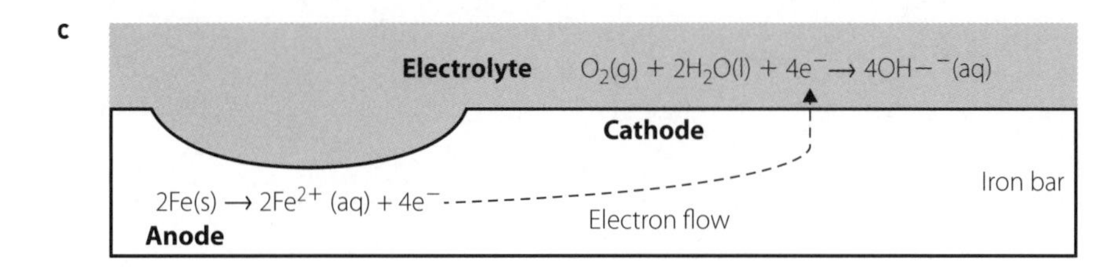

4 a

b Elements left out: D, G, K, M, N

c These elements have been left out of the activity series because they are noble gases and are extremely unreactive.

d It would be difficult to rank the activities of elements 22–29 because these elements have very similar ionisation energies. Therefore, more information is required to rank these elements.

5

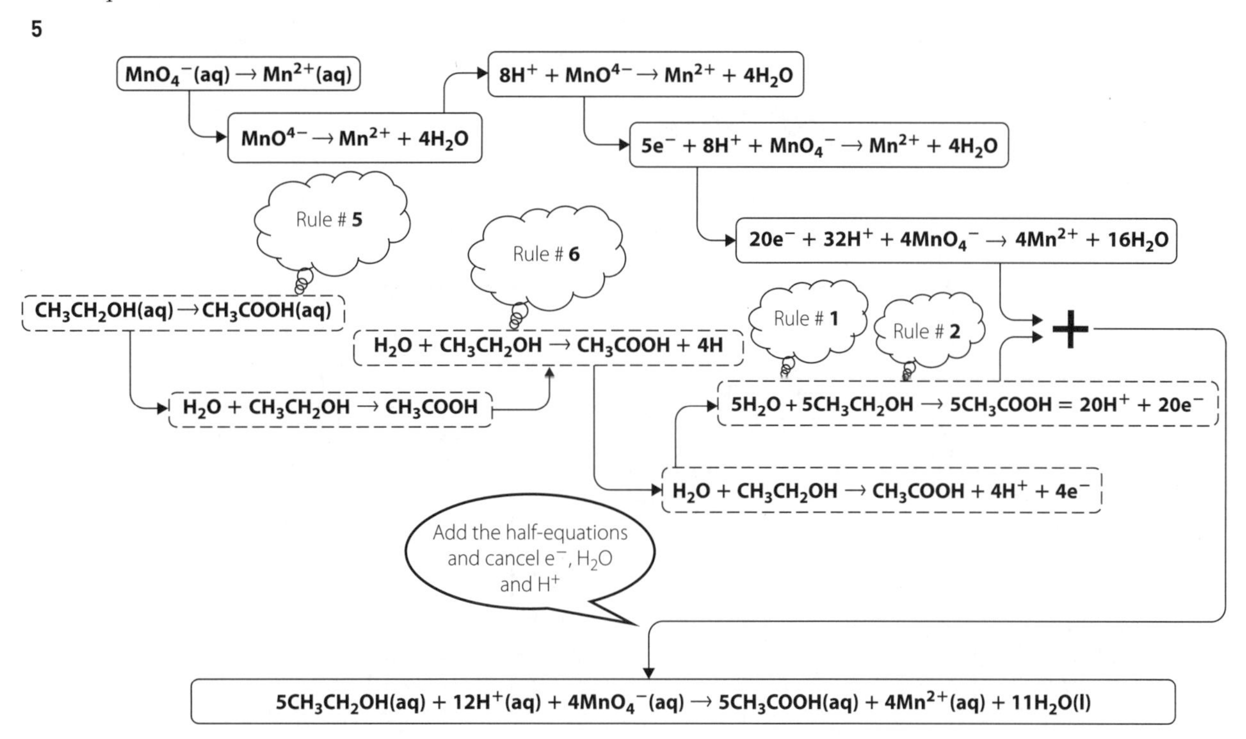

6 $C_2O_4^{2-} \rightarrow CO_2$

$4 \times -2 = -8$ $\quad$ $2 \times -2 = -4$

$-8 + C_2 = -2$ $\quad$ $C = +4$

$C_2 = +6$

$Cr = +3$

Carbon has been oxidised. This reaction *is* a redox reaction.

7 a

$3Ca(HCO_3)_2(aq) + 2H_3BO_3(aq) \rightarrow Ca_3(BO_3)_2(aq) + 6CO_2(g) + 6H_2O(l)$

$HCO_3^- = -1$ $\quad$ $BO_3^{3-} = -3$

$\therefore Ca = +2$ $\quad$ $\therefore Ca = +2$

Ca has neither been oxidised nor reduced. This is *not* a redox reaction.

b $NH_4Cl(aq) + NaOH(aq) \rightarrow NH_3(aq) + NaCl(aq) + H_2O(l)$

$N + (4x + 1) + (-1) = 0$ $\quad$ $N + (3x + 1) = 0$

$\therefore N = -3$ $\quad$ $\therefore N = -3$

N has neither been oxidised nor reduced. This is *not* a redox reaction.

c Note: Elemental oxygen = 0; oxidation state of oxygen = −2

$C_8H_{18}(l) + 25O_2(g) \rightarrow 16CO_2(g) + 18H_2O(l)$

$(8 \text{ x } C) + (18 \times 1) = 0$ $\quad$ $C + (2 \times -2) = 0$

$\therefore C = -2.25$ $\quad$ $\therefore C = +4$

EVALUATION

1 B **2** D **3** B **4** B

5 **a** +7

b +3

c +3

d 0

6 Number of protons in the nucleus, distance of the valence electrons from the nucleus, number of electrons shells between the valence shell and the nucleus

7 A is more reactive than L. Therefore, A is oxidised, L is reduced.

(Assume both metals react to produce 2+ ions.)

$A(s) \rightarrow A^{2+}(aq) + 2e^-$ Reducing species

$L^{2+}(aq) + 2e^- \rightarrow L(s)$ Oxidising species

8 $5e^- + 8H^+(aq) + MnO_4^-(aq) \rightarrow Mn^{2+}(aq) + 4H_2O(l)$

$C_2O_4^{2-}(aq) \rightarrow 2CO_2(g) + 2e^-$

$16H^+(aq) + 2MnO_4^-(aq) + 5C_2O_4^{2-}(aq)$
$\rightarrow 2Mn^{2+}(aq) + 10CO_2(g) + 8H_2O(l)$

CHAPTER 11 ELECTROCHEMICAL CELLS

11.1 ELECTROLYTIC CELLS

1

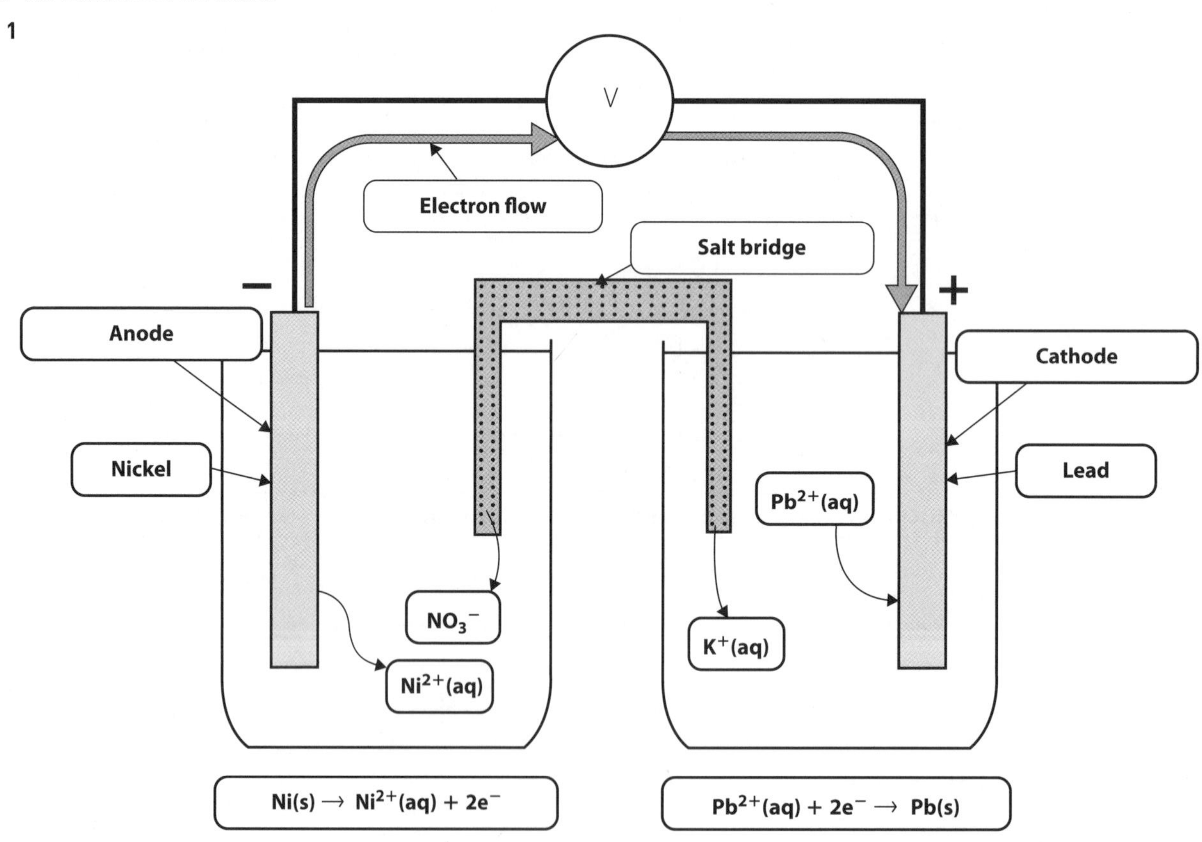

 9780170412476

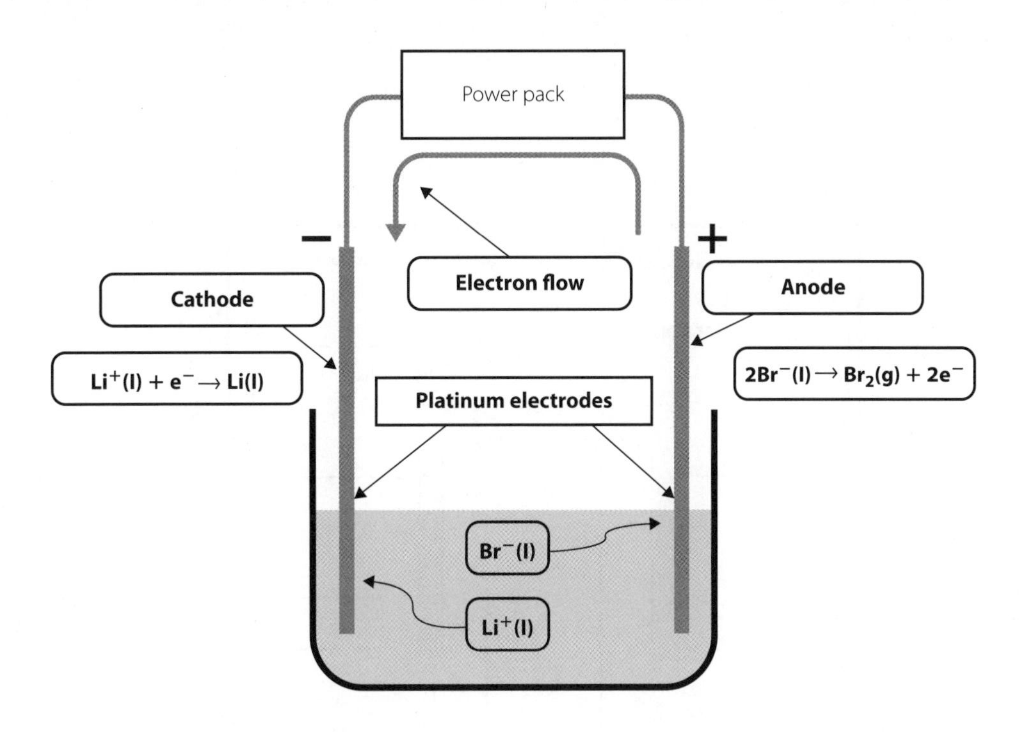

2 a

CELL	METAL 1	METAL 2
A	Cu	Zn
C	Mg	Zn
B	Pb	Fe
E	Sn	Cu
F	Mg	Pb
D	Ni	Mg

b A 1 < 2 Metal 2: $Zn(s) \rightarrow Zn^{2+}(aq) + 2e^-$

Metal 1: $Cu^{2+}(aq) + 2e^- \rightarrow Cu(s)$

B 1 < 2 Metal 2: $Fe(s) \rightarrow Fe^{2+}(aq) + 2e^-$

Metal 1: $Pb^{2+}(aq) + 2e^- \rightarrow Pb(s)$

C 1 > 2 Metal 1: $Mg(s) \rightarrow Mg^{2+}(aq) + 2e^-$

Metal 2: $Zn^{2+}(aq) + 2e^- \rightarrow Zn(s)$

D 2 < 1 Metal 2: $Mg(s) \rightarrow Mg^{2+}(aq) + 2e^-$

Metal 1: $Ni^{2+}(aq) + 2e^- \rightarrow Ni(s)$

E 1 > 2 Metal 1: $Sn(s) \rightarrow Sn^{2+}(aq) + 2e^-$

Metal 2: $Cu^{2+}(aq) + 2e^- \rightarrow Cu(s)$

F 1 > 2 Metal 1: $Mg(s) \rightarrow Mg^{2+}(aq) + 2e^-$

Metal 2: $Pb^{2+}(aq) + 2e^- \rightarrow Pb(s)$

3

	GALVANIC CELL	ELECTROLYTIC CELL
Reaction spontaneity	**Spontaneous**	**Non-spontaneous**
Power source	**Chemical reaction**	**External**
Direction of electron flow	**Anode → Cathode**	**Anode → Cathode**
Anode charge	**Negative**	**Positive**
Cathode charge	**Positive**	**Negative**
Location of oxidation and reduction	**Oxidation at anode Reduction at cathode**	**Oxidation at anode Reduction at cathode**
Number of cells and why	**Two half-cells to separate the reactants**	**One cell to contain compound to be electrolysed**
Electrode polarity determined by	**Determined by the reaction occurring at each electrode**	**The connection of the electrodes to the power source**

EVALUATION

1 D **2** C **3** D **4** B

5 For the reaction:

$A(s) + B^{2+}(aq) \rightarrow A^{2+}(aq) + B(s)$

At the anode: Oxidation $A(s) \rightarrow A^{2+}(aq) + 2e^-$

At the cathode: Reduction $B^{2+}(aq) + 2e^- \rightarrow B(s)$

The cathode is metal B which is the *reduced species* produced by the reduction of $B^{2+}(aq)$.

6 $Mg^{2+}(l) + 2e^- \rightarrow Mg(l)$ Reduction

$2Br^-(l) \rightarrow Br_2(g) + 2e^-$ Oxidation

7

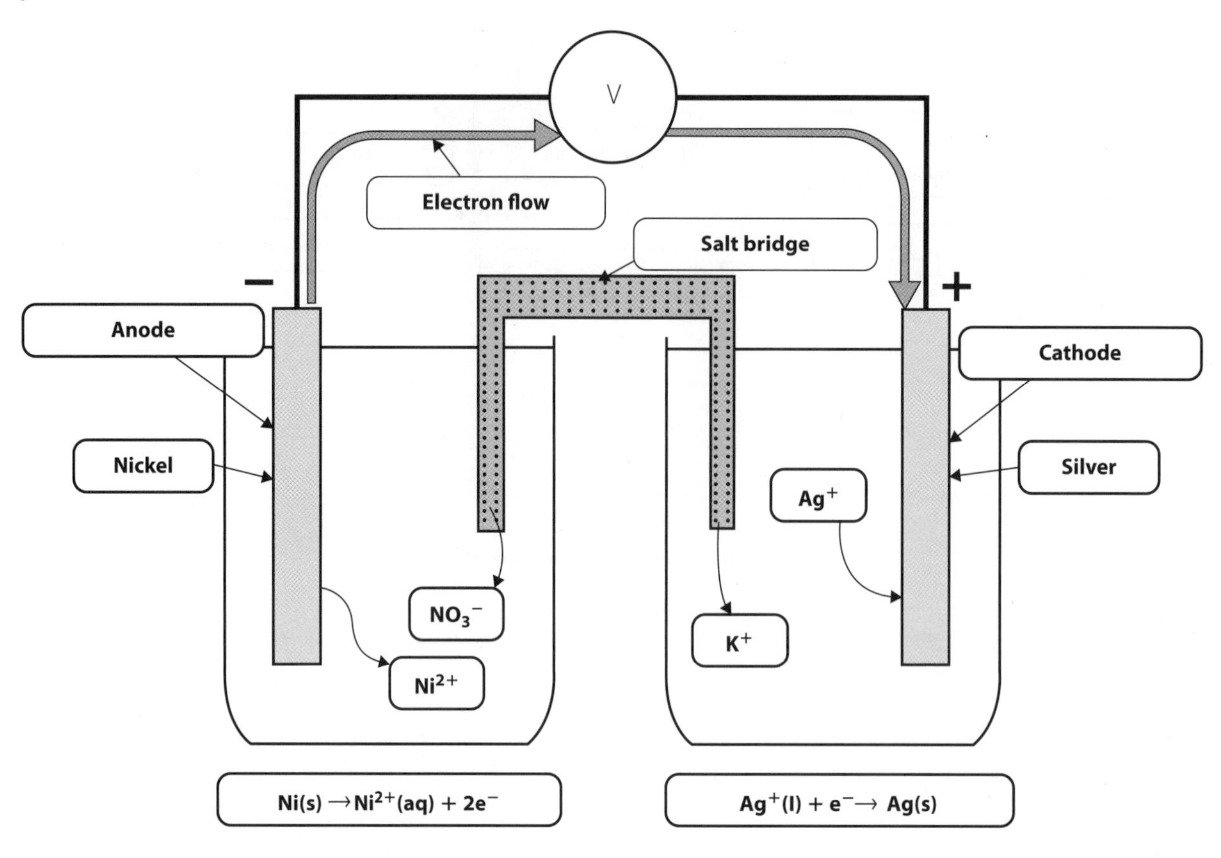

CHAPTER 12 GALVANIC CELLS

12.1 GALVANIC CELLS

1

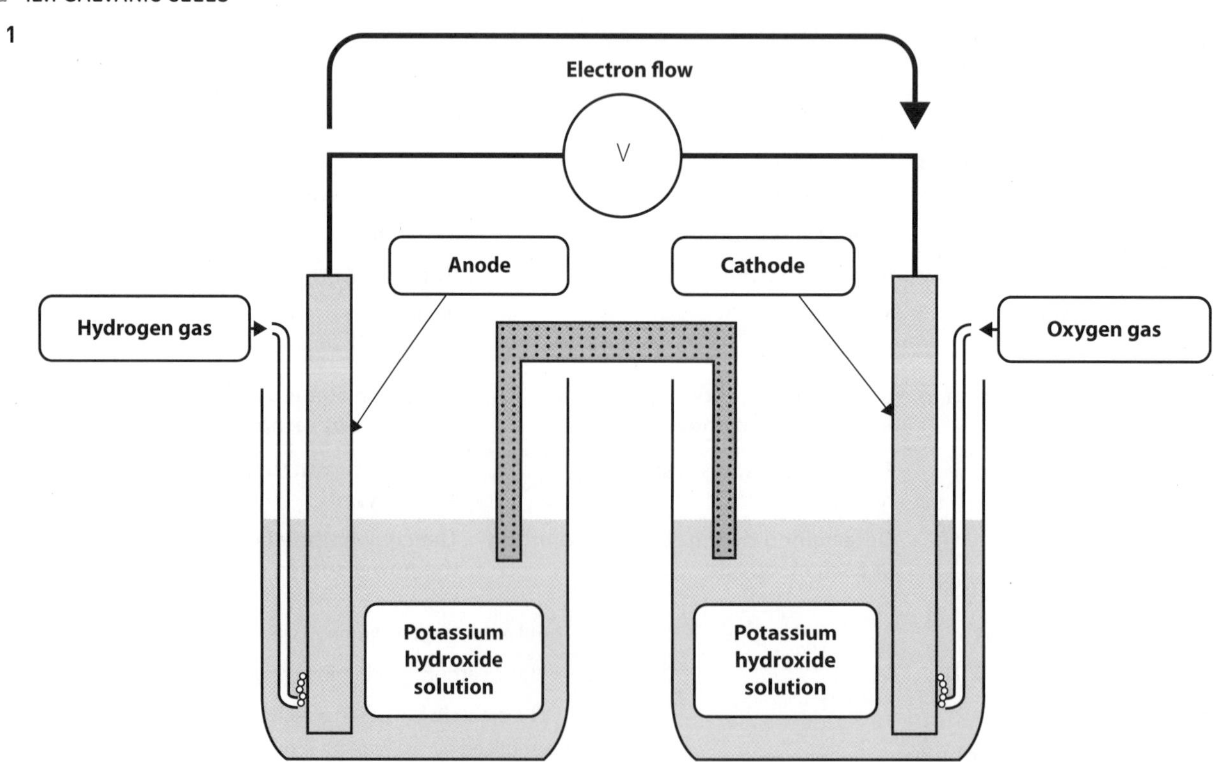

2

	ELECTRODE X	ELECTRODE Y
A	Zn $Zn \rightarrow Zn^{2+} + 2e^-$ Oxidation	Cu $Cu^{2+} + 2e^- \rightarrow Cu$ Reduction
B	Pb $Pb^{2+} + 2e^- \rightarrow Pb$ Reduction	Zn $Zn \rightarrow Zn^{2+} + 2e^-$ Oxidation
C	Cu $Cu^{2+} + 2e^- \rightarrow Cu$ Reduction	Fe $Fe \rightarrow Fe^{2+} + 2e^-$ Oxidation
D	Fe $Fe \rightarrow Fe^{2+} + 2e^-$ Oxidation	Pb $Pb^{2+} + 2e^- \rightarrow Pb$ Reduction

EVALUATION

1 D **2** C **3** C **4** D

5 Sodium nitrate and potassium nitrate do not react with any of the electrodes or electrolytes.

6 Anode = oxidation

$Sn(s) \rightarrow Sn^{2+}(aq) + 2e^-$

Cathode = Reduction

$Cu^{2+}(aq) + 2e^- \rightarrow Cu(s)$

This solid copper attaches to the copper cathode, which, as a result, gains mass.

7 Anode: $H_2(g) + 4OH^-(aq) \rightarrow 4H_2O(l)$

Cathode: $O_2(g) + 2H_2O(l) \rightarrow 4OH^-(aq)$

Both electrodes are lined with a platinum mesh. The platinum acts as a catalyst for the breaking of the gaseous H—H and O—O bonds.

8 During operation of the galvanic cell, the electrolyte solutions become more dilute. This decreases the rate of reaction that produces electricity.

CHAPTER 13 STANDARD ELECTRODE POTENTIAL

13.1 ELECTROCHEMICAL SERIES

1 ab

	Oxidised form ne^- → Reduced form	$E°$(V)
Strongest oxidising agent	$F_2 + 2e^- \rightarrow 2F^-$	+2.87
	$Au^+ + e^- \rightarrow Au$	+1.69
	$Cl_2 + 2e^- \rightarrow 2Cl^-$	+1.36
	$Br_2 + 2e^- \rightarrow 2Br^-$	+1.09
	$Cu^{2+} + 2e^- \rightarrow Cu$	+0.34
	$2H^+ + 2e^- \rightarrow H_2$	0.00
	$Pb^{2+} + 2e^- \rightarrow Pb$	−0.13
	$Ni^{2+} + 2e^- \rightarrow Ni$	−0.26
	$Cd^{2+} + 2e^- \rightarrow Cd$	−0.40
	$Fe^{2+} + 2e^- \rightarrow Fe$	−0.45
	$Zn^{2+} + 2e^- \rightarrow Zn$	−0.76
	$Mg^{2+} + 2e^- \rightarrow Mg$	−2.37
	$Na^+ + e^- \rightarrow Na$	−2.71
Strongest reducing agent	$K^+ + e^- \rightarrow K$	−2.93

c The half-equations are all written as reduction reactions to enable easy comparison of the relative strengths of the substances as oxidising or reducing agents.

2

OXIDISED FORM $2e^- \rightarrow$ REDUCED FORM	$E°$ (V)
$L^{2+}(aq) + 2e^- \rightarrow L(s)$	−0.32
$Q^{2+}(aq) + 2e^- \rightarrow Q(s)$	+0.78
$T^{2+}(aq) + 2e^- \rightarrow T(s)$	−0.80
$X^{2+}(aq) + 2e^- \rightarrow X(s)$	−0.08
$R^{2+}(aq) + 2e^- \rightarrow R(s)$	+0.29
$G^{2+}(aq) + 2e^- \rightarrow G(s)$	−0.42

13.2 DETERMINING STANDARD CELL POTENTIALS

1 a

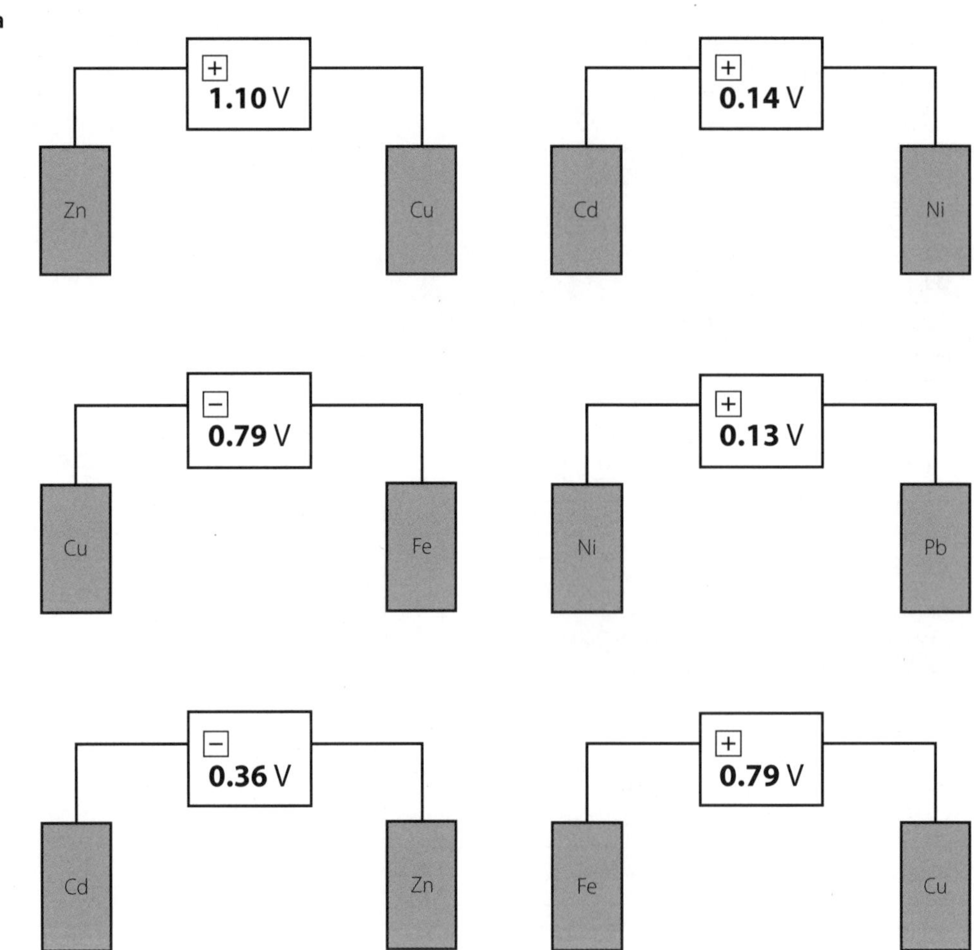

b Reactions that have a value for cell potential are non-spontaneous. These are Cu/Fe and Cd/Zn.

EVALUATION

1 D **2** C **3** B **4** A

5 Al should displace Cu from Cu^{2+} since it is a more active metal. However, Al reacts with oxygen in the air to form an unreactive coating of Al_2O_3 on its surface.

6 $Q^{2+} + 2e^- \rightarrow Q \; E° = -0.62\,V$

$T^{2+} + 2e^- \rightarrow T \; E° = +0.38\,V$

Q is the more active metal and so would lose electrons more easily. This would be the cathode and T would be the anode.

$Q \rightarrow Q^{2+} + 2e^-$ Oxidation = cathode

$T^{2+} + 2e^- \rightarrow T$ Reduction = anode

7 $E°$ values give an indication of the difference in electrical potential energy between the two sides of the equation. This does NOT change when the quantities in the equation change.

9780170412476

CHAPTER 14 ELECTROLYTIC CELLS

14.1 ELECTROLYSIS OF MOLTEN SALTS

1

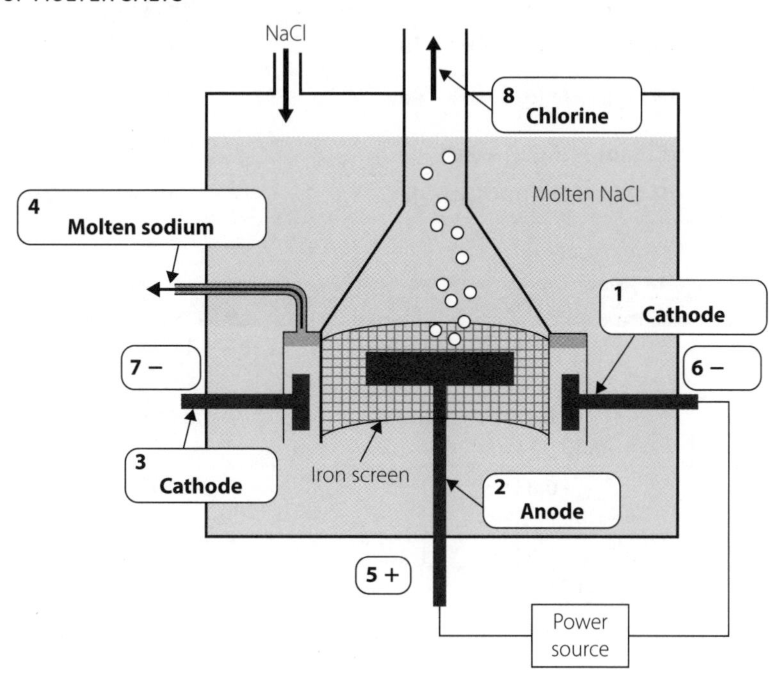

14.2 ELECTROLYSIS OF AQUEOUS SOLUTIONS

1 ab

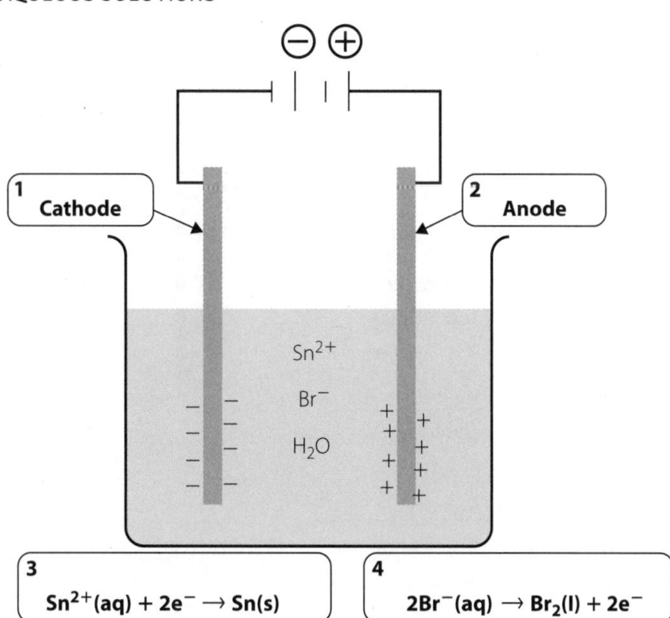

2 a X = Anode, Y = Cathode

b

LETTER	NUMBER	ANODE HALF-EQUATION	CATHODE HALF-EQUATION
A	4	$2Cl^-(aq) \rightarrow Cl_2(g) + 2e^-$	$Cu^{2+}(aq) + 2e^- \rightarrow Cu(s)$
B	2	$2H_2O(l) \rightarrow O_2(g) + 4H^+ + 4e^-$	$Cu^{2+}(aq) + 2e^- \rightarrow Cu(s)$
C	5	$2H_2O(l) \rightarrow O_2(g) + 4H^+ + 4e^-$	$H_2O(l) + 2e^- \rightarrow H_2(g) + 2OH^-(aq)$
D	1	$2Cl^-(aq) \rightarrow Cl_2(g) + 2e^-$	$H_2O(l) + 2e^- \rightarrow H_2(g) + 2OH^-(aq)$
E	3	$2H_2O(l) \rightarrow O_2(g) + 4H^+ + 4e^-$	$Zn^{2+} + 2e^- \rightarrow Zn(s)$

c

LETTER	NUMBER	CALCULATION	VOLTAGE REQUIRED
A	4	$E^\circ_{cell} = E^\circ_{red} + 0.34 + E^\circ_{ox} - 1.36$ $= -1.02$	1.02 V
B	2	$E^\circ_{cell} = E^\circ_{red} + 0.34 + E^\circ_{ox} - 1.23$ $= -0.89$	0.89 V
C	5	$E^\circ_{cell} = E^\circ_{red} - 0.83 + E^\circ_{ox} - 1.23$ $= -2.06$	2.06 V
D	1	$E^\circ_{cell} = E^\circ_{red} - 0.83 + E^\circ_{ox} - 1.36$ $= -2.19$	2.19 V
E	3	$E^\circ_{cell} = E^\circ_{red} - 0.76 + E^\circ_{ox} - 1.23$ $= -1.99$	1.99 V

EVALUATION

1 D **2** C **3** A **4** B

5 At the anode the two possible products are oxygen and chlorine. The E° values for the oxidation of chloride ions and the oxidation of water are very similar. There are more chloride ions present in a concentrated chloride solution and thus there is a greater probability of oxidation of chloride ions to chlorine gas.

6 In addition to sodium, the process occurring in the Downs cell produces large amounts of chlorine gas – an industrially and commercially important gas.

7 $2H_2O(l) \rightarrow O_2(g) + 4H^+ + 4e^-$ $E^\circ = -1.23\,V$ Oxidation

$H_2O(l) + 2e^- \rightarrow H_2(g) + 2OH^-(aq)$ $E^\circ = -0.83\,V$ Reduction

$E^\circ_{cell} = -0.83 + (-1.23) = -2.06\,V$

The electrolysis of water requires a minimum of 2.06 V. This is larger than the 1.23 V produced by the alkali fuel cell and, therefore, the closed loop of energy transformations won't work.

8 ab

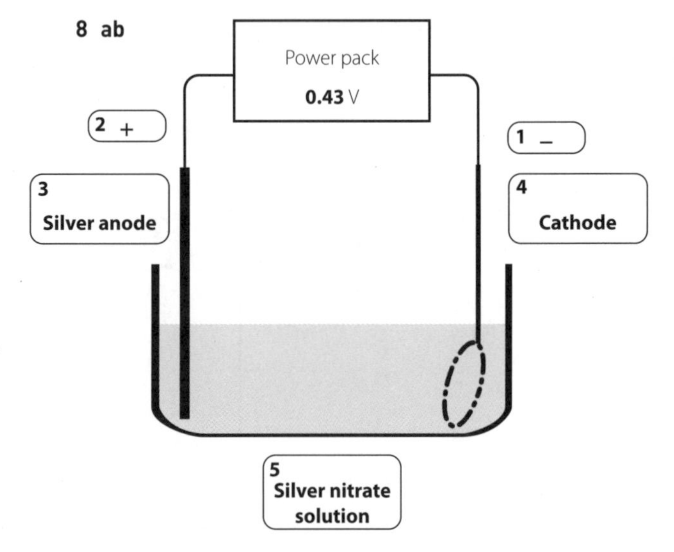

c $2H_2O(l) \rightarrow O_2(g) + 4H^+ + 4e^-$ $E^\circ = -1.23\,V$

$Ag^+(aq) + e^- \rightarrow Ag(s)$ $E^\circ = +1.69\,V$

$E^\circ_{cell} = +0.80 + (-1.23) = -0.43\,V$

9780170412476

CHAPTER 15 STRUCTURE OF ORGANIC COMPOUNDS

15.1 IMPORTANT TERMS

1 Crossword answers

Across		Down	
2	Prefix	1	Catenate
4	Structural	3	Functional
6	Skeleton	5	Stem
8	Homologous	7	Molecular
10	Geometric	9	Suffix
11	Semistructural		

2 Answers will vary.

15.2 DRAWING STRUCTURAL AND SEMI-STRUCTURAL FORMULAS

1 **a** $CH_3(CH)_2CH_2CH_3$

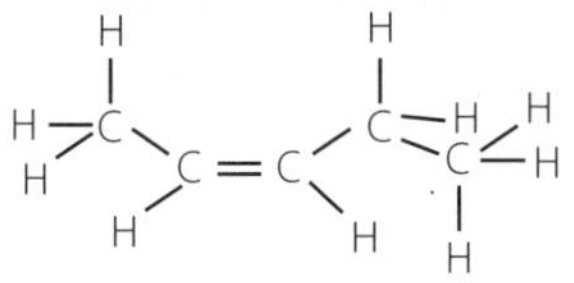

b $CH_3CHCH_2CH_2Br$

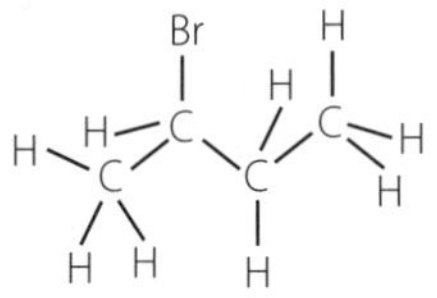

c $CH_3(CH)_2CH_2CHO$

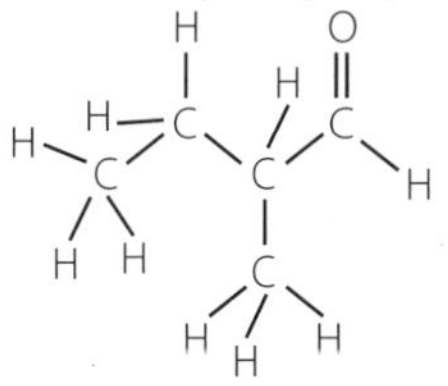

d $CH_3CH_2COCH_2CH_3$

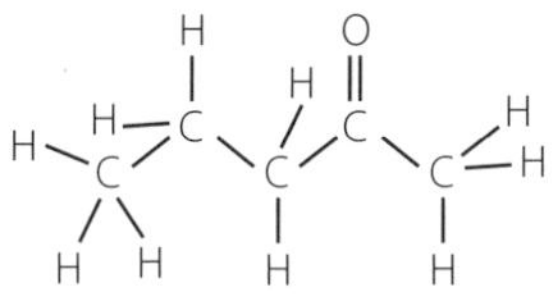

e CH_3CH_2COOH

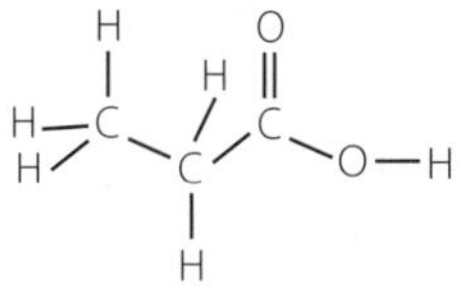

f $CH_3CH_2COOCH_2CH_3$

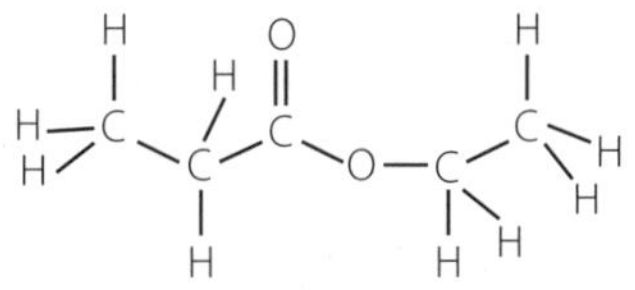

2 **a** It should be 2-butanol.

b 2,2-dichloro-1-pentene cannot exist. If a C=C double bond is attached to C atom number 2 it cannot also be bonded to two chlorine atoms.

3 Yes.

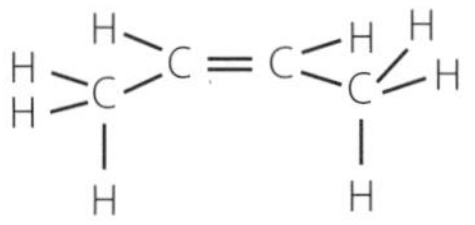

cis-2-butene

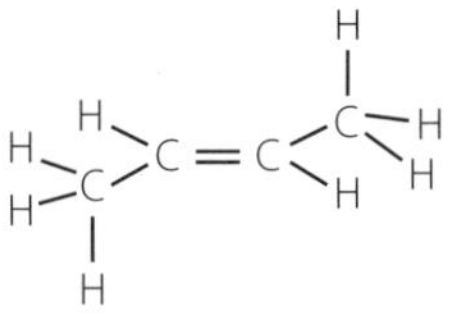

trans-2-butene

4 No, the two bromine atoms on C atom 1 mean that no geometric isomerism is possible.

15.3 NAMING MOLECULES AND IDENTIFYING FUNCTIONAL GROUPS

1

Molecule 1	a **Saturated hydrocarbon** b **Alkane** c **3,4-dimethylhexane** d **No**
Molecule 2	a **C=C bond** b **Alkene** c **3-methyl-2-hexene** d **Yes**
Molecule 3	a **Hydroxyl group** b **Alcohol** c **2-pentanol** d **No**
Molecule 4	a **Hydroxyl group** b **Alcohol** c **2-pentanol** d **No**
Molecule 5	a **Carboxyl group** b **Carboxylic acid** c **Hexanoic acid** d **No**

15.4 GEOMETRIC ISOMERISM

1 **a** These molecules are identical molecules of one substance. They are not geometric isomers.

b These molecules are identical molecules of one substance. They are not geometric isomers.

c These are geometric isomers: one is trans-1,2-dichloroethene and the other is cis-1,2-dichloroethene.

d These molecules are both cis-1,2-dichloroethene.

EVALUATION

1 **a** A series of molecules with the same functional group but different length carbon chain.

b Isomers are two molecules with the same molecular formula but different structural formulas. Alkenes display geometric isomerism, where, because of the lack of rotation of the molecule around a C=C double bond, two possible arrangements of the atoms in space are possible for molecules with the same structural formula.

c

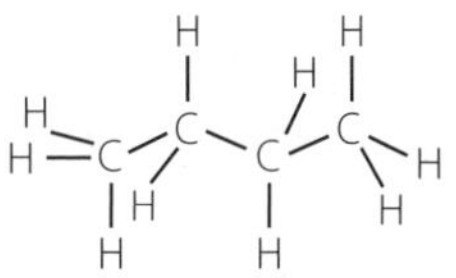

Butane

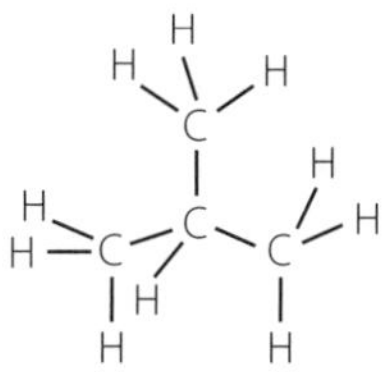

2-methylpropane

d

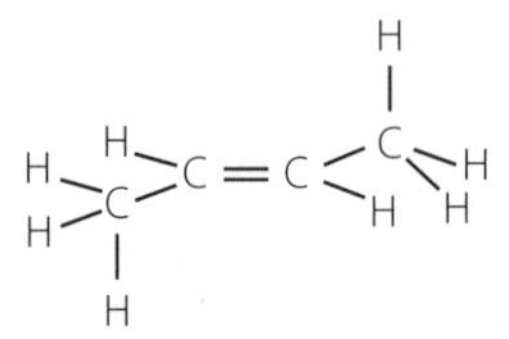

trans-2-butene

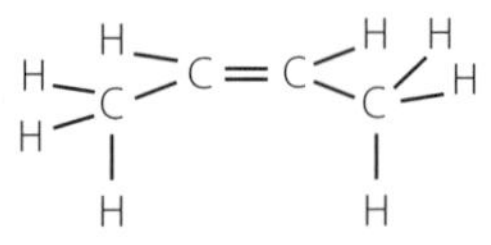

cis-2-butene

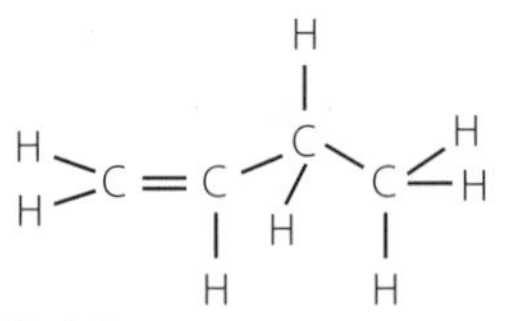

But-1-ene

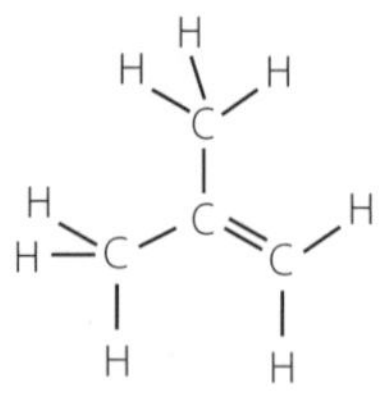

2-methylpropene

CHAPTER 16 PHYSICAL PROPERTIES AND TRENDS OF ORGANIC MOLECULES

16.1 IMPORTANT TERMS

1 Crossword answers

Across		Down	
3	Electronegativity	**1**	Hydrophobic
6	Hydrophilic	**2**	Intermolecular
7	Dispersion	**4**	Dipole
		6	Hydrogen
		7	Non-polar

2 Answers will vary.

16.2 COMPARE AND CONTRAST

1 a Propanoic acid is more soluble as it has a shorter hydrophobic alkyl chain.

b 1,2-pentanediol is more soluble, as it has two hydroxyl groups which form hydrogen bonds with water molecules.

c Methyl methanoate is more soluble, as it has a smaller hydrophobic alkyl chain.

2 No. Glycerol a highly polar molecule with three hydroxyl groups and would not be soluble in the non-polar hexane.

3 a Heptene has the higher boiling point. It is a larger molecule and, therefore, has greater dispersion forces acting between molecules.

b 2-pentanol has the higher boiling point. It has intermolecular hydrogen bonding whereas the non-polar butane does not.

c 1,2,3-propanetriol has the greater boiling point. It has a greater intermolecular hydrogen bonding due to the higher number of hydroxyl groups.

16.3 SOLUBILITY OF ALCOHOLS

1 The greater the carbon chain length, the lower is the solubility of the molecule. As the carbon chain length increases, the greater is the proportion of the molecule which is non-polar and, therefore, will not form positive intermolecular bonds with water molecules.

2 When comparing molecules with the same length carbon chain, carboxylic acid molecules have a greater solubility than alcohol molecules. When added to water, the acid molecules ionise and form ion–dipole bonds with water molecules, in addition to hydrogen bonding interactions that exist between alcohol molecules and water.

3 Chloroalkanes would be less soluble than alcohols, as the C–Cl bond is polar but much less so than the C–O or the O–H bond in alcohols. Consequently, the intermolecular bonds that form are weaker.

Alkanes would be much less soluble than alcohols – they are non-polar molecules, and therefore form no dipole-dipole forces with water molecules. They would be immiscible with water.

Amines would have a similar solubility as alcohols. They have an N–H bond, which has a similar polarity to the O–H bond, and therefore would form hydrogen bonds of similar strength to that of alcohols. Clearly, amines would become less soluble with increasing chain length.

EVALUATION

1 C

2 a As the number of carbon atoms increases, the molecular size increases and, therefore, the capacity for forming intermolecular dispersion forces also increases as they act between the surfaces of the molecules. The energy required to separate molecules in the liquid state also increases.

b Intermolecular forces are greater between alcohols than between the corresponding alkane molecules due to the presence of the hydroxyl groups which cause hydrogen bonding to occur between the molecules.

c As the molecules get larger, the proportion of the molecule that is hydrophobic increases compared to that which is hydrophilic. Therefore, the ability to form hydrogen bonds is reduced and the additional effect of the hydroxyl groups in alcohols is less significant.

9780170412476

CHAPTER 17 ORGANIC REACTIONS AND REACTION PATHWAYS

17.1 IMPORTANT TERMS

1 Crossword answers

Across		Down	
6	Dehydration	**1**	Reduction
9	Radical	**2**	Hydrolysis
13	Incomplete	**3**	Monoprotic
16	Ester	**4**	Addition
18	Primary	**5**	Secondary
19	Oxidation	**7**	Elimination
20	Fractional	**8**	Halogenation
		10	Condensation
		11	Tertiary
		12	Complete
		14	Nucleophilic
		15	Esterification
		17	Hydrogenation

2 Answers will vary.

17.2 REACTION TYPES

	DESCRIPTION	EXAMPLE, WITH REAGENTS AND CONDITIONS
Addition	**Two molecules combining to form one large molecule**	**Ethene + hydrogen → ethane (hydrogen with a nickel catalyst at 350°C)**
Combustion	**A molecule such as hydrocarbon reacting in oxygen, forming carbon dioxide and water**	**Methane + oxygen → carbon dioxide + water**
Condensation	**Two molecules combining to produce a larger molecule, with the production of water as a by-product**	**Ethanol + propanoic acid → ethyl propanoate + water (concentrated sulfuric acid catalyst)**
Elimination	**A larger molecule that reacts to form two smaller molecules**	**Ethanol → ethene + water (catalyst is phosphoric acid at 180°C)**
Oxidation	**Conversion of primary alcohols to aldehydes or carboxylic acids; or secondary alcohols to ketones**	**1-propanol → propanoic acid (reacting with acidified potassium dichromate)**
Reduction	**Addition of hydrogen to either alkenes or nitriles**	**1-propane nitrile → 1-butanamine (hydrogen with a nickel catalyst at 350°C)**
Substitution (free radical)	**Replacement of an atom on an organic molecule with another, such as a halogen, in the presence of UV light**	**Ethane + chlorine → chloroethane + hydrogen chloride (UV light)**
Substitution (nucleophilic)	**Reaction of a haloalkane, where the halogen atom is replaced by another functional group such as a hydroxyl or amino group**	**Chloropropane + aqueous sodium hydroxide → sodium chloride + 1-propanol**

17.3 DRAWING AND IDENTIFYING STRUCTURES AND REACTIONS

1 a Primary alcohol

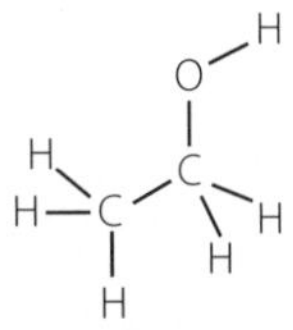

H H O H C C H H H

b Secondary alcohol

O H H H C C C H H H H H

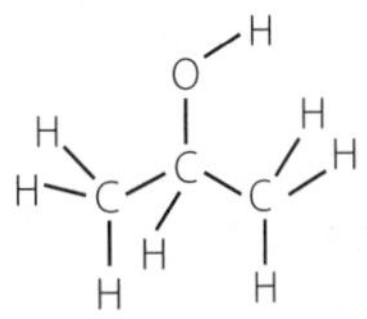

c Tertiary alcohol

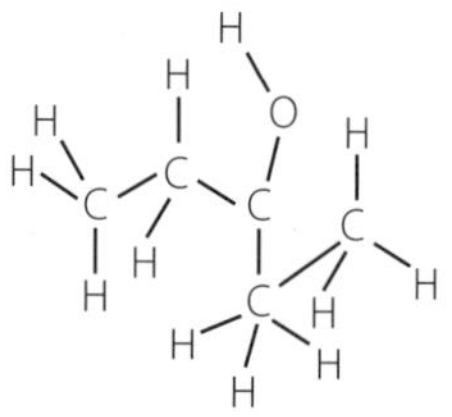

H
H
H O H
H C C C C H
H H
H C H
H H
H

2 a Ethanol

b 2-propanol

c 2-methyl-2-butanol

3 With addition of acidified potassium dichromate: ethanol would be oxidised to ethanal and then ethanoic acid, 2-propanol would be oxidised to propanone, and 2-methyl-2-butanol would not react.

4

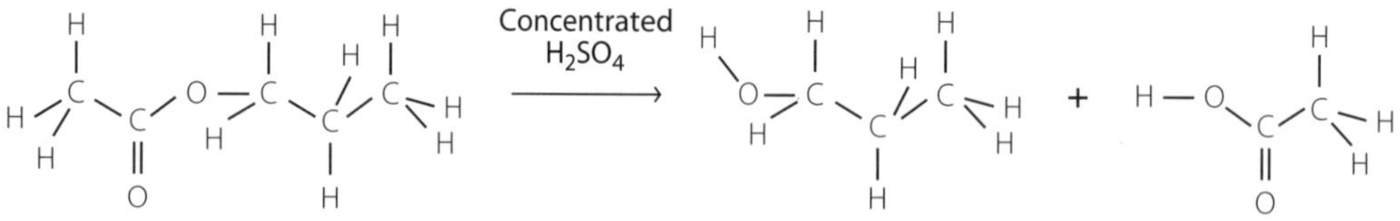

Concentrated H_2SO_4 →

Propyl ethanoate

Propanol

Ethanoic acid

Propyl ethanoate would be hydrolysed to ethanoic acid and propanol.

5 a Condensation reaction

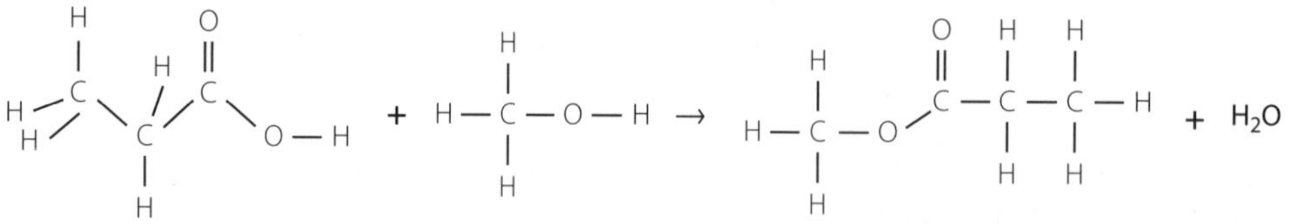

Propanoic acid + Methanol → Methyl propanoate + H_2O

Propanoic acid

Methanol

Methyl propanoate

Water

b Addition reaction

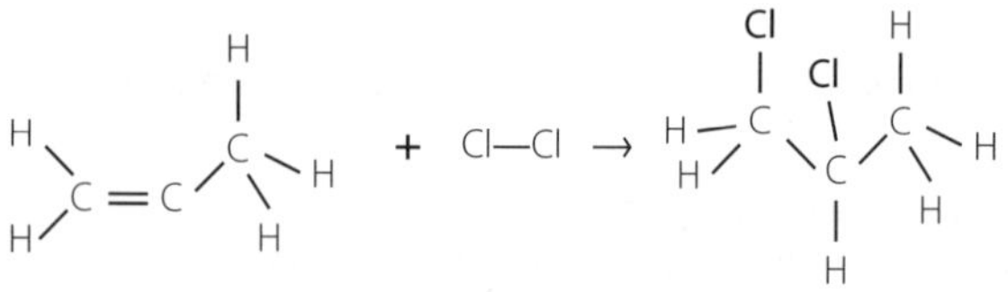

+ Cl—Cl →

1-propene

Chlorine

1,2-dichloropropane

c Substitution reaction

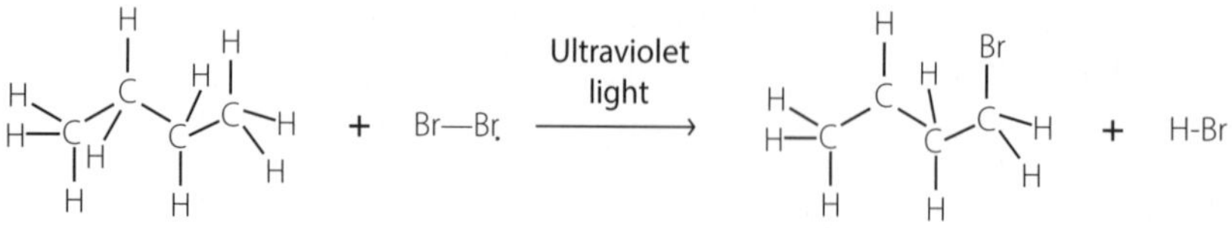

+ Br—Br Ultraviolet light → + H-Br

Butane

Bromine

1-bromobutane

Hydrogen bromide

17.4 IDENTIFYING REACTIONS

1 Samples of reactions are given below. Other answers are possible.

a Free radical substitution with Br_2 and ultraviolet light

9780170412476

b Addition reaction

3,4-dimethylhexane + $Br—Br$ $\xrightarrow{\text{Ultraviolet light}}$ 1-bromo-3,4-dimethylhexane + $H—Br$

3,4-dimethylhexane | Bromine | 1-bromo-3, 4-dimethylhexane | Hydrogen bromide

c Oxidation

3-methyl-2-hexene + $Cl—Cl$ → 2,3-dichloro-3-methylhexane

3-methyl-2-hexene | Chlorine | 2,3-dichloro-3-methylhexane

d Elimination

2-pentanol $\xrightarrow{\text{Acidified } K_2Cr_2O_7}$ 2-pentanone

2-pentanol | 2-pentanone

e Esterification

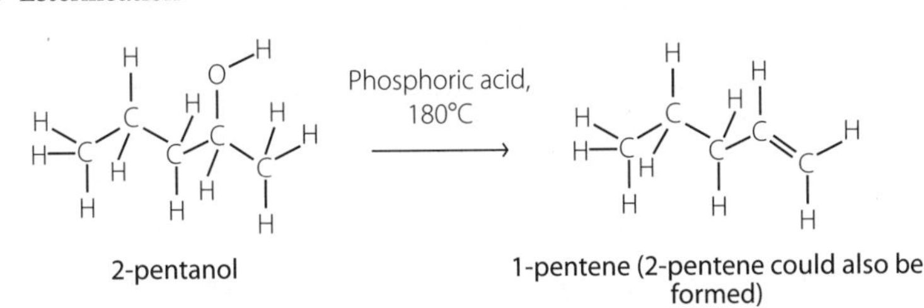

2-pentanol | 1-pentene (2-pentene could also be formed)

EVALUATION

Hexanoic acid + Methanol $\xrightarrow{c.H_2SO_4}$ Methyl hexanoate + H_2O

Hexanoic acid | Methanol | Methyl hexanoate | Water

1 **A** $CH_3CH_2CH_2CH_2OH$

B $CH_3CHOHCH_3$

2 **i** Cl_2, Ultraviolet light

ii Acidified $K_2Cr_2O_7$

iii NaOH (aq)

iv Concentrated H_2SO_4

3 Step II Nucleophilic substitution

Step III Oxidation

Step IV Free radical substitution

Step V Nucleophilic substitution

Step VI Esterification/condensation

4 Step III Butanoic acid

Step IV 2-propanol

5 Step II NaCl

Step IV HCl

Step VI Water

CHAPTER 18 ORGANIC MATERIALS: STRUCTURE AND FUNCTION

18.1 IMPORTANT TERMS

1 Crossword answers

Across		Down	
3	Copolymer	1	Polyamide
4	Amino	2	Polymerisation
5	Density	4	Cross-links
8	Protein	6	Branching
9	Polymer	7	Condensation
12	Tensile	8	Polypeptide
15	Zwitterion	10	Quaternary
16	Side	11	Addition
17	Biodegradable	13	Secondary
		14	Monomer
		18	Isoelectric

2 Answers will vary.

18.2 DOMINOES

Beginning All living things are massive chemical machines composed largely of organic compounds interacting in a watery soup.

Polymers are composed of giant molecules formed from small molecules (monomers) joined by covalent bonds.

Natural polymers that are proteins include silk, wool and hair.

The characteristic of thermoplastic polymers is that they soften on heating and can be moulded.

Polymers that do not soften as the temperature is raised are called thermosetting polymers.

Addition polymerisation occurs without loss of atoms from the monomers.

The monomers from which addition polymers are formed have a double bond between adjacent carbon atoms.

What are the monomer units in polyethylene? Ethene (ethylene) molecules, $CH_2{=}CH_2$.

Low-density polyethylene has weak intermolecular forces because chain branching prevents close packing of the molecules.

Because of the absence of branching in HDPE, the packing of molecules is regular, and called crystalline.

What is the arrangement of side chains in isotactic linear polymers? They are all on the same side of the polymer chain.

Because of the random distribution of side chains in atactic polymers, attraction between chains is low and the polymers are soft.

Non-polymer molecules that reduce attractions between chains are called plasticisers.

Polymers that have more than one monomer in the chains are called copolymers.

Cross-linking that joins chains together gives rise to two- or three-dimensional structures that are more rigid.

Condensation polymers are formed from monomers that have two different functional groups.

From monomers with both a carboxylic acid and an alcohol, polyesters are formed.

Monomers with both a carboxylic acid and an amine form polyamides in which the monomer units are joined by the amide link, –CO–NH–.

The monomer units in the carbohydrates cellulose and starch are molecules of glucose, a monosaccharide.

Sucrose, maltose and lactose are disaccharides.

Because disaccharide molecules are too big to pass through cell membranes they must be broken down to the monosaccharides for digestion.

The monomer units of proteins are amino acids.

Both the carboxylate and the amine groups on amino acids are protonated if they are in low pH conditions.

The primary structure of a protein is the sequence of amino acids along the protein chain.

Hydrogen bonding between amino acids on the same of different chains gives rise to the secondary structure of proteins.

The tertiary structure of a protein is the overall shape, resulting in the folding of the helices or sheets of the secondary structure.

Biofuels are regarded as renewable energy sources because they can be continually produced from crops or algae.

The production of ethanol from sugars, in the presence of yeasts, is called fermentation. **End**

18.3 EXPLAINING PROPERTIES OF FATS AND OILS

1 Triglycerides in fats have higher melting points; they are solid at room temperature.

2 Saturated molecules have a more flexible structure; therefore, they will pack together more easily than unsaturated molecules.

3 Stearic acid, because saturated molecules will pack more closely together.

4 Stearic acid would have more surface area overlap and, therefore, more surface for intermolecular attraction to occur.

5 Stearic acid would be the fat.

6 Hydrogenation of the hydrocarbon tails will saturate the molecules and, therefore, turn liquid oils into fats, such as margarine, which is more useful as a solid.

9780170412476

18.4 COMPARING AND CONTRASTING BIOLOGICAL POLYMERS

1

NAME	MONOMER UNIT(S)	NAME OF BOND LINKING MONOMERS	BIOLOGICAL ACTIVITY	DESCRIPTION OF STRUCTURE
Proteins	**Amino acid**	**Peptide**	**Used for growth and repair**	**Primary structure of polypeptide chains held together by intermolecular interactions to form secondary, tertiary and quaternary structures**
Saturated lipids	**Saturated fatty acids**	**Ester link**	**Forms solid fats**	**Triglyceride ester with saturated carbon chains**
Unsaturated lipids	**Unsaturated fatty acids**	**Ester link**	**Forms oils**	**Triglyceride ester with unsaturated carbon chains**
Cellulose	**β-glucose**	**Glycosidic**	**Forms plant structures**	**Unbranched polymers but with significant cross-linking to form a rigid structure**
Starch (amylose)	**α-glucose**	**Glycosidic**	**Energy store of energy for plants**	**Straight polymer chains**
Starch (amylopectin)	**α-glucose**	**Glycosidic**	**Energy store of energy for plants**	**Highly branched polymer chains**

EVALUATION

1 a The tertiary structure is the three-dimensional structure formed by a single peptide as a result of the accumulated intermolecular interactions between functional groups in the amino acid monomers. This is very important for the biological function of the protein.

b Denaturation can occur either through exposure to high temperature or to extremes of pH. Both of these factors can cause permanent changes to the tertiary structure of the protein by affecting the intermolecular bonding between the peptide chains and, therefore, the way in which the chains interact with each other. Therefore, the properties of the protein are permanently changed (denatured) also.

c The structure of enzymes has developed through evolution; they are constructed with a specifically defined tertiary or quaternary structure and held together by the intermolecular interactions. The sheer number of intermolecular interactions and orientation of the peptide chains leads to a complexity which cannot be replicated in a laboratory. As such, synthetic catalysts do not operate with the level of efficiency that natural enzymes do.

2 a

H OH HO OH H H H OH O OH

β-glucose

b Glycosidic

c Cellulose is made from α-glucose, which is not made or metabolised by animals. However, starch is made from β-glucose, which animal enzymes can metabolise.

3 a A: H_2SO_4

B: Glycerol–$CH_2OHCHOHCH_2OH$

b The fatty acid product has the formula $C_{17}H_{31}COOH$. This indicates that it has two C=C bonds and, therefore, can be described as polyunsaturated.

CHAPTER 19 ANALYTICAL TECHNIQUES

19.1 IMPORTANT TERMS

1 Crossword answers

Across		Down	
3	Interference	**1**	Destructive
4	Radical	**2**	Diffraction
8	Fraction	**5**	Cation
9	Electrophoresis	**6**	Buffer
12	Incidence	**7**	Parent
14	Chromatography	**10**	Anion
		11	Stationary
		13	Electric
		15	Mobile
		16	Adsorb

2 Answers will vary.

19.2 IR SPECTROSCOPY AND BOND VIBRATIONS

Answers will vary depending on the options chosen and molecules studied.

19.3 ISOTOPES AND MASS SPECTROMETRY

1 a −9

b −6

c −1

2 The abundance ratio of ^{35}Cl to ^{37}Cl is 3:1. Therefore, the likely abundance of ^{35}Cl–^{35}Cl:^{35}Cl–^{37}Cl:^{37}Cl–^{37}Cl is 9:6:1, as demonstrated by the following table, which shows the possible combinations:

	^{35}Cl	^{35}Cl	^{35}Cl	^{37}Cl
^{35}Cl	70	70	70	72
^{35}Cl	70	70	70	72
^{35}Cl	70	70	70	72
^{37}Cl	72	72	72	74

3

	^{12}C (98.9%)	^{13}C (1.1%)
^{16}O (99.76%)	^{12}C–^{16}O (RMM 28)	^{13}C–^{16}O (RMM 29)
^{18}O (0.20%)	^{12}C–^{18}O (RMM 30)	^{13}C–^{18}O (RMM 31)

A CO molecule with a RMM of 28 will be the most abundant.

4 a $^{12}C^{16}O$, $^{14}N_2$ and $^{12}C_2{}^{1}H_4$

b They all have an integral relative mass of 28.

c $CO = 12.00000 + 15.99491$

$= 27.99491$

$N_2 = 14.00307 \times 2$

$= 28.00614$

$C_2H_4 = (2 \times 12.00000) + (4 \times 1.00783)$

$= 28.031$

d The molecule is N_2.

EVALUATION

1 a The amino acids separate into different layers in the columns because the different molecules adsorb by different amounts to the stationary phase and dissolve in different quantities in the mobile phase. Therefore, they move at different rates through the column and are separated into the different bands shown.

b

BAND NUMBER	AMINO ACID
1	**Arginine**
2	**Glutamic acid**
3	**Cysteine**
4	**Proline**
5	**Valine**

2 a

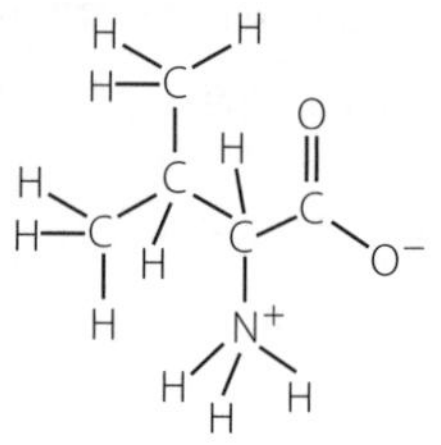

b

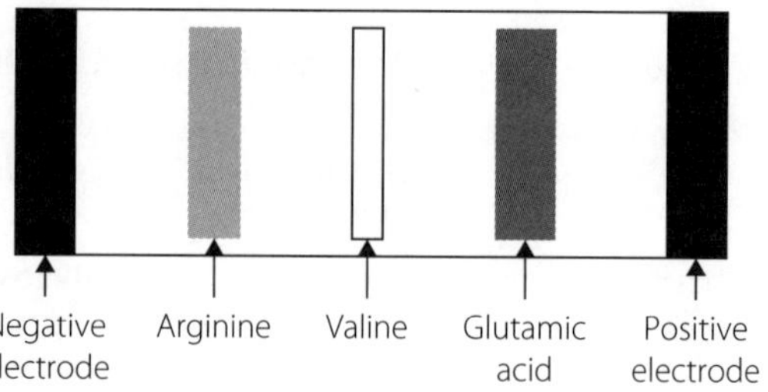

Negative electrode | Arginine | Valine | Glutamic acid | Positive electrode

3 a The IR spectrum of ethanol would show a significant, broad absorption at approximately 3500 cm^{-1} due to the O—H bond that is present. This absorption would not be present in dimethylether.

b Due to the presence of the O—H bond, the mass spectrum would show the presence of a peak at 17 *m/z* which would not be present in dimethyl ether. Due to its symmetrical nature, dimethyl ether would show major peaks at 31 and 15 *m/z*, because these fragments are formed due to the breaking of two possible bonds.

4 The peaks are caused by the fragmentation of the parent molecule into the following ions, each of which has a different mass: 17–OH^+, 31–CH_2OH^+, 61–$CH_2OHCHOH^+$ and 92–$CH_2OHCHOHCH_2OH^+$ (parent ion).

5 Bond lengths can be determined from the wavelengths of the X-rays being used when diffraction occurs. Bond angles can be deduced from analysing the position of atoms from multiple X-ray diffraction patterns taken as the crystal being analysed is rotated through many different positions.

6 Acetone would have a strong, sharp absorption at approximately 1750 cm^{-1} (O–H). Prop-2-ene-1-ol would have two significant absorptions, one at approximately 1660 cm^{-1} (C=C) and one large broad absorption at approximately 3500 cm^{-1} (O–H).

9780170412476

CHAPTER 20 CHEMICAL SYNTHESIS

20.1 IMPORTANT TERMS

1 Crossword answers

Across		Down	
2	Excess	1	Mechanism
8	Intermediate	3	Contact
9	Transesterification	4	Yield
11	Haber	5	Limiting
13	Biofuel	6	Pathway
		7	Synthetic
		10	Cracking
		12	Fuel

2 Answers will vary.

20.2 THEORETICAL AND PERCENTAGE YIELD

1 a F^- is the limiting reagent.

b Theoretical yield is 0.001 moles of CaF_2.

2 a $n(\text{methanol}) = \frac{1000}{32}$

$= 31.25$

$n(H_2) = 2 \times 31.25$

$= 62.5$ moles

$m(H_2) = 2 \times 62.5$

$= 125\,g\ H_2$

b $\%\ \text{yield} = \frac{103.5}{125} \times 100$

$= 82.8\%$

20.3 RAW MATERIALS

1 Nitrogen gas, hydrogen gas

2 Because the readily available nitrogen in the air can be 'fixed' and converted into nitrate and ammonium compounds that can form fertilisers and therefore help to grow food.

3 Mass production of fertilisers is now possible. Previously, nitrogen fixation was only possible through the growing of legume crops.

4 Developing a process that was economic and efficient: producing high enough yields of ammonia at a reasonable rate of reaction and at a sufficiently low cost.

5 Answers will vary, depending on research conducted and argument presented.

20.4 EXTENT OF REACTION AND TEMPERATURE

1 a 0.00500 moles N_2 and 0.0125 moles H_2

b 3.18×10^{-5} moles of NH_3 present at equilibrium

c i 4.77×10^{-5} moles

ii 4.95×10^{-3} moles

iii 4.95×10^{-3} M

d i 1.59×10^{-5} moles

ii 0.0109 moles

iii 0.0109 M

e

	$3N_2$	H_2	$\leftarrow\rightarrow$	$2NH_3$
At start	0.00500	0.0125		0
	$-(3.18 \times 10^{-5}) \times \frac{3}{2} = -4.77 \times 10^{-5}$	$-\frac{(3.18 \times 10^{-5})}{2} = -1.59 \times 10^{-5}$		$+3.18 \times 10^{-5}$
At equilibrium	4.95×10^{-3}	0.0109		3.18×10^{-5}

$$K = \frac{[NH_3]^2}{[N_2]^3[H_2]}$$

$$= \frac{(3.18 \times 10^{-5})^2}{(4.95 \times 10^{-3})^3 \times 0.0109}$$

$$= 0.765\ M^{-2}$$

2 $\%$ of N_2 reacted $= \frac{4.95 \times 10^{-3}}{0.005} \times 100$

$= 99.0\%$

3 $\%$ of H_2 reacted $= \frac{0.0109}{0.0125} \times 100$

$= 87.2\%$

4 The amount would be greater, it's an exothermic reaction; therefore, the forward reaction is favoured by a lower temperature.

20.5 THE ROLE OF CATALYSTS IN THE PRODUCTION OF MATERIALS FROM SYNGAS

1 Economic considerations: availability and cost of the different raw materials compared to coal

2 a Iron(III) oxide, aluminium oxide and cerium(IV) oxide

b Copper, zinc and aluminium of chromium oxides

c Iron carbide catalyst

20.6 REVISION QUESTIONS

1 Crude oil (e.g. octane, or petrol), nitrogen gas (e.g. ammonia), hydrogen gas (e.g. syngas or ammonia), coal (e.g. synga), sulfur (e.g. sulfuric acid).

2 Rate of production, equilibrium yield, cost of raw materials, cost of conditions (e.g. temperature and pressure), workplace health and safety, and environmental impact of waste production.

3 High rate and equilibrium yield at low temperature and pressure, high atom economy (e.g. little wastage), and minimal environmental impact of obtaining raw materials and disposing of waste.

4 a Low equilibrium yield of ammonia and low reactivity of nitrogen gas.

b Compromised temperature and pressure required.

c Use of catalyst to reduce optimum temperature required, compromise pressure used, recycling of unused nitrogen and hydrogen, and use of heat exchangers to utilise heat from exothermic reaction.

5 Compromise temperature and pressure utilised, and production of oleum to minimise heat loss from exothermic reaction.

6 Lowering or raising temperature depending on whether forward reaction is exothermic or endothermic, raising or lowering pressure according to whether volume decreases or increases, and removing products once formed so that reaction does not reach equilibrium.

7 Theoretical yield is the maximum amount of product that can be produced from a given amount of reactants; percentage yield is the percentage of this maximum amount of product that is actually produced, due to in efficiencies in the synthetic process.

8 The ability of enzymes to distinguish between enantiomers of glucose.

9 Different petroleum molecules have different demands as sources of chemical products and as fuels. Also, different samples of crude oil contain differing amounts of differently sized molecules, each with different uses.

EVALUATION

1 a The reaction is exothermic and raising the temperature reduces the equilibrium yield; however, the reaction occurs at a lower rate at lower temperature. Therefore, a compromise temperature is chosen of 300°C, together with a catalyst, which produces an acceptable equilibrium yield of ethanol at an acceptably high rate of reaction.

b Higher pressure leads to a higher yield because there is a decrease in volume across the reaction and the yield of products is favoured. However, high pressure is expensive to achieve; therefore, a compromise pressure is used which achieves an acceptable yield at an acceptable cost.

2 a Compound A is methanol CH_3OH.

b Compound B is glycerol $C_3H_8O_3$.

c $CH_3CHCH(CH_2)_{14}COOCH_3$

CHAPTER 21 GREEN CHEMISTRY

21.1 IMPORTANT TERMS

1 Crossword answers

Across	Down
2 Economy	1 Pollution
3 Synthetic	4 Hazardous
7 Green	5 Solvent
8 Efficient	6 Renewable
9 Catalyst	

2 Answers will vary.

21.2 CALCULATING ATOM ECONOMY

1 a $\%\ \text{economy} = \frac{32}{34} \times 100$

$= 94.1\%$

b Energy and environmental costs of obtaining raw materials

c 100%

2 100%, there are no other products.

3 $\%\ \text{economy} = \frac{32}{(16 + 160 + 18)} = 16.5\%$

21.3 THE MANUFACTURE OF IBUPROFEN

1 Benzene ring and carboxyl

2 $\%\ \text{yield} = 0.8 \times 0.8 \times 0.8 \times 0.8 \times 0.8 \times 100$

$= 32.8\%$

3 $\frac{1500}{40} \times 100$

$= 3750$ tonnes waste product

4 $\%\ \text{yield} = 0.8 \times 0.8 \times 100$

$= 64\%$

5 Use of catalysts enables the reaction pathway to occur with fewer steps, as reactions that previously could not occur without a catalyst are now possible. Therefore, fewer waste molecules are produced and the atom economy is significantly improved.

EVALUATION

1 C

2 a CO_2 should not be released into the environment and instead can be used for a different purpose (e.g. carbonating drinks).

b 1 Prevent waste: 100% of products can be re-used. Caffeine extracted is sold to drug companies and coffee beans used to make coffee.

2 Design less hazardous processes: No toxic chemicals used in this process.

3 Increase energy efficiency: Supercritical CO_2 is a much more efficient solvent and reduces the quantity of chemicals and energy required to achieve the same outcome.

9780170412476

CHAPTER 22 MACROMOLECULES: POLYMERS, PROTEINS AND CARBOHYDRATES

22.1 ADDITION POLYMERISATION

1 Polypropene

2

```
 C                               C         C
 |                               |         |
—C — C — C — C — C — C — C — C — C —
          |         |
          C         C
```

Atactic

```
 C              C              C
 |              |              |
—C — C — C — C — C — C — C — C — C —
          |              |
          C              C
```

SYNtactic

```
—C — C — C — C — C — C — C — C — C —
 |         |         |         |         |
 C         C         C         C         C
```

ISOtactic

3 $CH_2=CH$

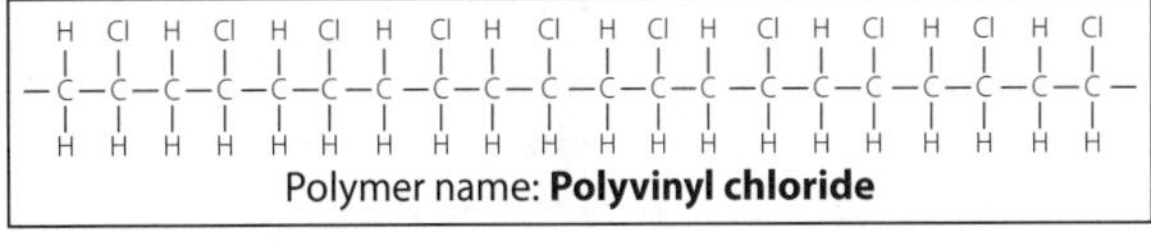

Polymer name: **Polyvinyl chloride**

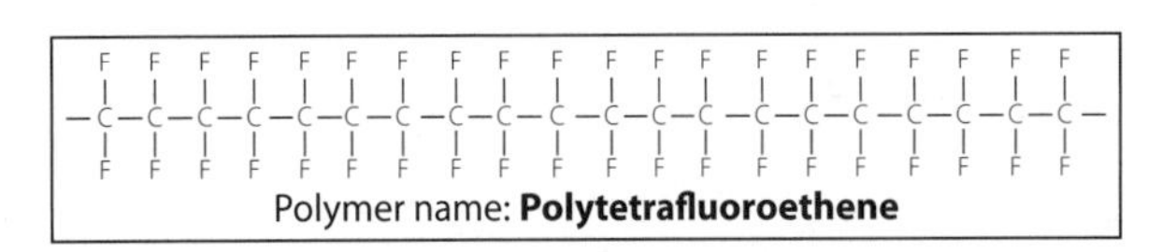

Polymer name: **Polytetrafluoroethene**

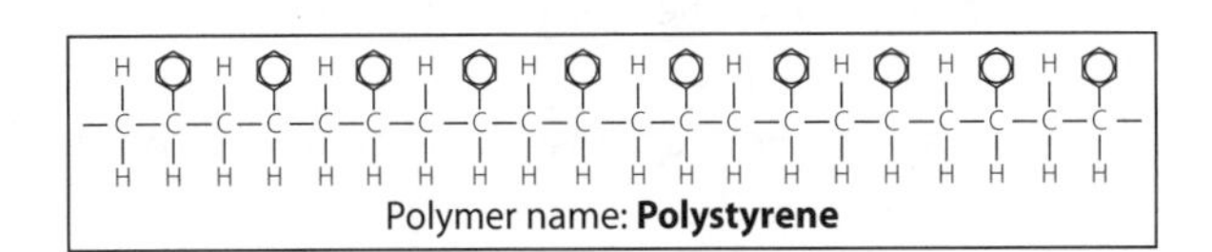

Polymer name: **Polystyrene**

```
   OOCCH3  OOCCH3  OOCCH3  OOCCH3  OOCCH3  OOCCH3  OOCCH3  OOCCH3  OOCCH3  OOCCH3
     |       |       |       |       |       |       |       |       |       |
—H2C—C—H2C—C—H2C—C—H2C—C—H2C—C—H2C—C—H2C—C—H2C—C—H2C—C—H2C—C—
     |       |       |       |       |       |       |       |       |       |
    CH3     CH3     CH3     CH3     CH3     CH3     CH3     CH3     CH3     CH3
```

Polymer name: **Polyvinyl acetate**

22.2 CONDENSATION POLYMERS

1

```
   H       O   H       O   H       O   H       O   H       O   H
   |       ||  |       ||  |       ||  |       ||  |       ||  |
H—C—O—C—C—O—C—C—O—C—C—O—C—C—O—C—C—H
   |           |           |           |           |           |
   H           H           H           H           H           H
```

Polyglycolic acid

2

```
              O         O                      O         O                      O         O                      O         O                      O         O
              ||        ||                     ||        ||                     ||        ||                     ||        ||                     ||        ||
O—CH2—CH2—O—C—⟨◯⟩—C—O—CH2—CH2—O—C—⟨◯⟩—C—O—CH2—CH2—O—C—⟨◯⟩—C—O—CH2—CH2—O—C—⟨◯⟩—C—O—CH2—CH2—O—C—⟨◯⟩—C—
```

22.3 CARBOHYDRATES

1 and **2**

H_2O

Glycosidic link

—C—O—C—

3

	CELLULOSE	STARCH	GLYCOGEN
Function	**Structural**	**Energy source**	**Energy source**
Where they are formed	**Plant cells**	**Chloroplasts in plants**	**Liver**
How they occur	**Fibrous structure**	**Grains**	**Small granules**
Monomer	**β-glucose**	**α-glucose**	**α-glucose**
How the polysaccharide is arranged	**Straight, long unbranched chains form H-bonds with nearby chains**	**Most starch is amylopectin consisting of long, branched chains**	**Highly branched with an amorphous structure**

22.4 PROTEINS

1ab

Primary structure **The sequence of the amino acids in a protein chain.**
Secondary structure **The way that amino acids in a protein chain bond to amino acids in the same or nearby chains.**
Tertiary structure **The three-dimensional shape of a protein.**
Quaternary structure **The three-dimensional structure consisting of the aggregation of two or more individual polypeptide chains that operate as a single unit.**

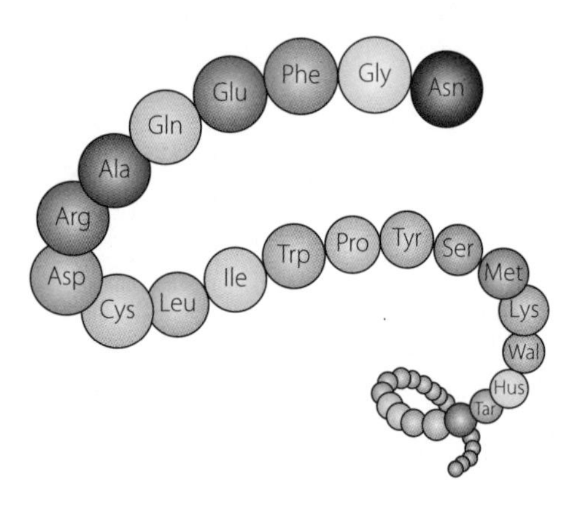

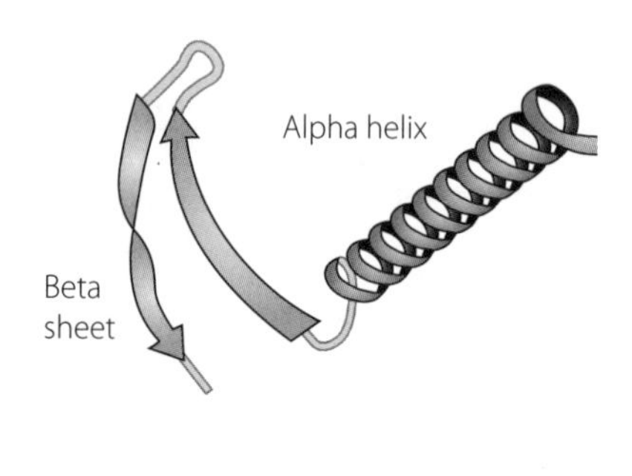

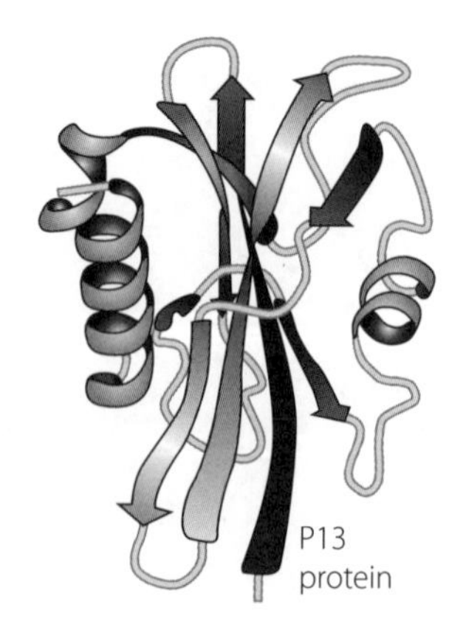

22.5 AMINO ACIDS

1

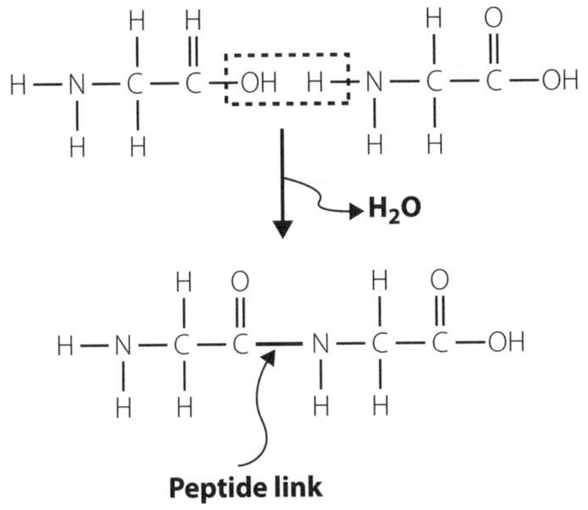

H_2O

Peptide link

2

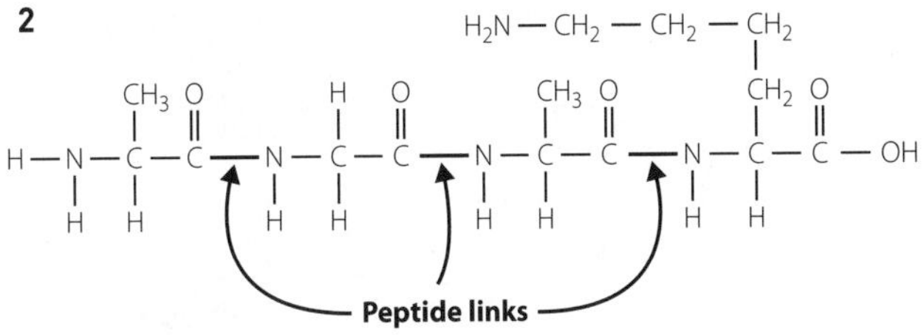

Peptide links

EVALUATION

1 B **2** B **3** D **4** C

5

POLYMER	STRUCTURE AND BONDING	PROPERTIES
Low-density polyethene (LDPE)	Amorphous (disordered) Branched chains	Soft, flexible, transparent, impermeable to water vapour, unreactive towards acids and bases, absorbs oils and softens, low melting point (80–95°C), low tensile strength
High-density polyethene (HDPE)	Crystalline (ordered) Linear chains	Denser, tougher, more rigid, higher melting point and greater tensile strength than LDPE, opaque, impermeable to water vapour, unreactive towards acids and bases

6 The Cl atoms in PVC are attached to the polymer chain to form strong dipole–dipole attractions with nearby PVC chains, giving a hard, rigid polymer.

The benzene ring attached to the polymer chains in polystyrene are non-polar, resulting in weak attractions between chains. The large rings prevent the chains from packing close together, giving a hard but brittle polymer.

7 Atactic: Randomly arranged side groups prevent close packing of chains polymers that are soft and with a low melting point.

Syntactic: Regular arrangement ensures chains can pack quite closely together resulting in a tough, clear polymer.

Isotactic: Arrangement of side groups on one side if the chain ensures that the chains can pack very close together, maximising intermolecular forces and resulting in a strong, hard polymer.

CHAPTER 23 MOLECULAR MANUFACTURING

23.1 THE ORIENTATION EFFECT

1 a

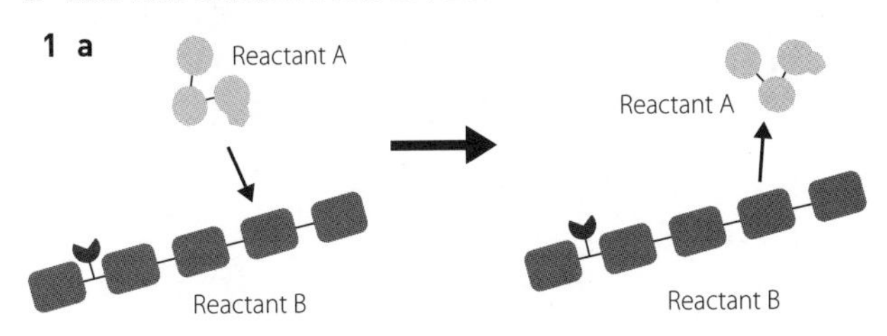

b

Unreactive protecting groups

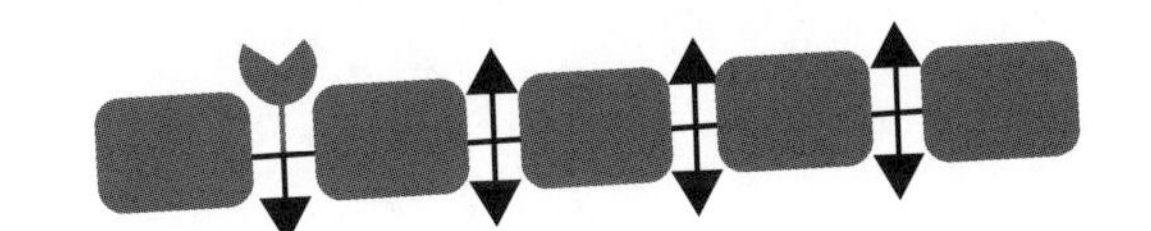

23.2 TOP-DOWN, BOTTOM-UP APPROACHES

1

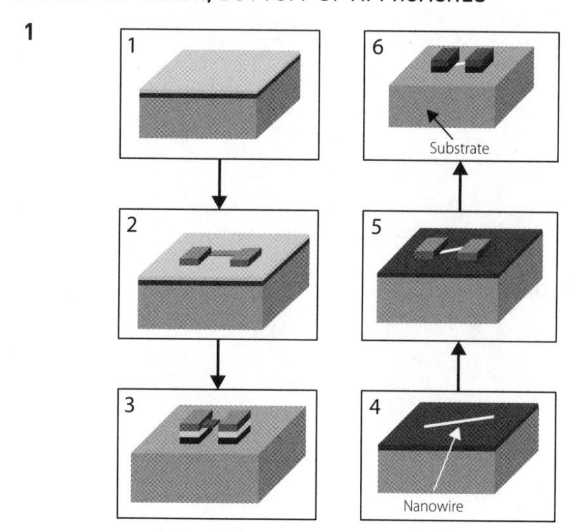

23.3 SYNTHETIC PROTEINS

1

NUMBER	IDENTIFYING LETTER
1	c
2	d
3	f
4	b

EVALUATION

1 C **2** A **3** B **4** A

5 L- and D-isomers are stereoisomers. They are mirror images of each other.

6 When propanone molecules in the bloodstream encounter the microcantilever array, they bond to the receptors on the microcantilevers. If the concentration of propanone reaches the danger level of about 0.7 ppm, the microcantilever will produce a voltage of about 4.2×10^{-6} V, which sets off an alarm to notify the person with diabetes that their blood sugar is too low.

9780170412476

PRACTICE EXAMINATION ANSWERS

MULTIPLE CHOICE

1 D	**2** C	**3** A	**4** A
5 B	**6** C	**7** C	**8** C
9 A	**10** D	**11** C	**12** D
13 B	**14** A	**15** B	**16** C
17 D	**18** A	**19** C	**20** C

SHORT ANSWER AND COMBINATION-RESPONSE QUESTIONS

1 a

Change to Reaction I and Reaction II	Effect of the change on the chlorine yield in Reaction I	Effect of the change on the chlorine yield in Reaction II
Increase in volume of the reaction vessel	**Increase**	**No change**
Reduction in temperature	**Decrease**	**Decrease**
Increase in concentration of PCl_3 (Reaction I) or H_2 (Reaction II)	**Decrease**	**Decrease**

b Reaction I: As the forward reaction occurs, there is an increase in the number of gaseous particles. When the reaction vessel volume increases, this causes a decrease in the total pressure in the vessel. Le Chatelier's principle indicates that the system acts to oppose this change by increasing the number of gaseous particles present. Hence, the equilibrium moves to the right and the forward reaction is favoured, increasing the equilibrium concentration of chlorine.

Reaction II: There is no change in volume in reaction II, and therefore the position of equilibrium is unaffected by a change in the reaction vessel volume.

c The enthalpy change for Reaction I is positive, meaning that the forward reaction is endothermic and the reverse reaction exothermic. Hence, the activation energy for the forward reaction is significantly larger than for the reverse reaction. Therefore, if the temperature is reduced, the rate of both forward and reverse reactions will fall. However, the rate of the forward reaction will fall by more than that of the reverse reaction, due to the larger activation energy. Therefore, the position of equilibrium will move to the left, and the equilibrium concentration of chlorine will fall.

d $c(PCl_5) = \dfrac{5.50}{20.0}$

$= 0.275\ mol\ L^{-1}$

$K = ([Cl_2] \times \dfrac{[PCl^3])}{[PCl^5]}$

$= 1.05$

As $[Cl_2] = [PCl_3]$, this becomes

$K = \dfrac{[Cl_2]^2}{[PCl^5]}$

$\doteq 1.05$

2 a $CH_3CH_2COOH + H_2O \rightarrow CH_3CH_2COO^- + H_3O^+$

b $K_a = [H_3O^+] \times \dfrac{[CH_3CH_2COO^-]}{[CH_3CH_2COOH]}$

c i $c(\text{propanoic acid}) = \dfrac{0.6}{3}$

$= 0.200\ M$

$c(\text{propanoate}) = \dfrac{0.200}{3}$

$= 0.0667\ M$

$[H^+] = K_a \times \dfrac{[HA]}{[A^-]}$

$= 1.35 \times 10^{-5} \times \dfrac{0.200}{0.0667}$

$= 4.04 \times 10^{-5}$

$pH = -\log(4.04 \times 10^{-5})$

$= 4.39$

ii The pH of an acid only solution would be lower. When propanoate is added, it has the effect of sending the equilibrium to the left and thus reducing the concentration of H^+ ions, thus increasing the pH.

3 a $Na_2CO_3(aq) + 2HCl(aq) \rightarrow 2NaCl(aq) + H_2O(l) + CO_2(g)$

b $n(Na_2CO_3)$ in sample $= 0.1 \times 1.00$

$= 0.1$ moles

$n(HCl)$ added $= 0.9 \times 0.0111$

$= 0.00999$ moles

$n(Na_2CO_3)$ reacted $= \dfrac{0.00999}{2}$

$= 4.995 \times 10^{-3}$

$n(Na_2CO_3)$ remaining $= 0.1 - 4.995 \times 10^{-3}$

$= 0.095005$ moles

(Na_2CO_3) remaining $= \dfrac{0.095005}{1}$

$= 0.09501\ M$

c The calculated concentration of nitric acid will be higher than actual value. Because the concentration of the sodium carbonate is lower than written value, the volume used in the titration will be greater. Therefore, it will appear that the concentration of the nitric acid in the aliquot is greater than it actually is.

4 a $n(MnO_4^-)$ remaining $= \dfrac{n(Fe^{2+})}{5}$

$= 0.148 \times \dfrac{0.02585}{5}$

$= 7.6516 \times 10^{-4}$ moles

b $n(MnO_4^-)$ reacted with $SO_2 = n(MnO_4^-)$ initial – $n(MnO_4^-)$ remaining

$n(MnO_4^-)$ initial $= 0.1 \times 0.0302$

$= 3.020 \times 10^{-3}$ moles

Therefore, $n(MnO_4^-)$ reacted $= 3.02 \times 10^{-3} - 7.6516 \times 10^{-4}$

$= 2.255 \times 10^{-3}$ moles

c $n(SO_2)$ reacting $= \frac{5}{2} \times n(MnO_4^-)$

$= 5.638 \times 10^{-3}$ moles

$m(SO_2)$ reacting $= 5.638 \times 10^{-3} \times 64$

$= 0.3608$ g

$c(SO_2) = \frac{0.3608}{5}$

$= 0.07217 \text{ g m}^{-3}$

5 a

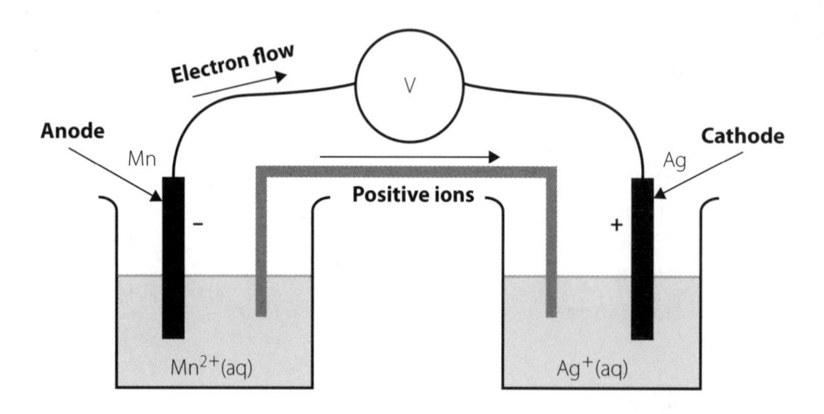

b Oxidation: $Mn(s) \rightarrow Mn^{2+}(aq) + 2e^-$

Reduction: $Ag^+(aq) + e^- \rightarrow Ag(s)$

Overall equation: $Mn(s) + 2Ag^+(aq) \rightarrow Mn^{2+}(aq) + 2Ag(s)$

c Substance for salt bridge needs to be: ionic, of high solubility and must not react with any of the other substances in the galvanic cell.

6 a $Cu^{2+}(aq) + 2e^- \rightarrow Cu(s)$

$Au^+(aq) + e^- \rightarrow Au(s)$

$Al^{3+} + 3e^- \rightarrow Al(s)$

b No. Moles deposited is proportional to the charge on the ion. Therefore, in cell order, the ratio is 2 : 1 : 3.

Therefore, n(Au) = 0.06 moles and n(Al) = 0.18 moles

7 a A $(CH_3)_2CH_2CH{=}CH_2$

B $(CH_3)_2CH_2CH_2CH_3$

C $(CH_3)_2CH_2CH_2CH_2Br$

D $(CH_3)_2CH_2CH_2CH_2OH$

E $(CH_3)_2CH_2CH_2CO_2H$

F $(CH_3)_2CH_2CHBrCH_3$

b X NaOH

Y $KMnO_4$ (or $K_2Cr_2O_7$)/ H^+

Z Concentrated H_2SO_4

c For example, $(CH_3)_2C{=}CH_2CH_3$

d **i** Addition

ii Substitution

iii Oxidation

8 a 8 double bonds: 3 in top section, 1 in middle section and 4 in bottom section

b 8 moles of iodine per mole of oil

Mass of 8 moles of iodine $= 8 \times 126.9$

$= 1015.2$

Iodine number $= \frac{\text{Mass iodine}}{\text{mass oil}} \times 100$

$= \frac{1015.2}{854} \times 100$

$= 118.9$

c Glycerol: $CH_2OHCHOHCH_2OH$

Fatty acid 1: $C_{18}H_{29}COOH$

Fatty acid 2: $C_{17}H_{31}COOH$

Fatty acid 3: $C_{15}H_{21}COOH$

EXTENDED RESPONSE QUESTION

Discussion should include, but not be limited to the following.

- Both polymers formed by condensation reactions: both have the same amide link. Proteins are formed in an enzyme catalysed reaction from amino acids, which contain an amino and a carboxyl group. Nylon is formed synthetically by reacting a diacid with a diester.
- Nylon fibres will be long and straight: much more uniform composition of identical monomer units. Proteins much more complex, less uniform structure due to a range of intermolecular interactions between the R-groups of the amino acid monomers. Proteins form enzyme-catalysed, very specific tertiary and quaternary structures. Due to the simplicity of its structure, nylon would be much less sensitive to variation in temperature or pH.
- Nylon is not biodegradable as there are no enzymes in nature that can digest it.